D1135313

Penguin Nature Guides

Fungi of Northern Europe · 2

Gill-Fungi

Sven Nilsson and Olle Persson
Illustrated by Bo Mossberg

Translated from the Swedish by David Rush
Edited and adapted by Dr David Pegler and
Brian Spooner

Penguin Books

Penguin Books Ltd, Harmondsworth,
Middlesex, England
Penguin Books, 625 Madison Avenue,
New York, New York 10022, U.S.A.
Penguin Books Australia Ltd, Ringwood,
Victoria, Australia
Penguin Books Canada Ltd, 2801 John Street,
Markham, Ontario, Canada L3R 1B4
Penguin Books (N.Z.) Ltd, 182–190 Wairau Road,
Auckland 10, New Zealand

Svampar i naturen 2 first published by
Wahlström & Widstrand 1977
This translation published 1978

Printed in Portugal by Gris Impressores
Filmset in Monophoto Times by
Northumberland Press Ltd,
Gateshead, Tyne and Wear

Contents

Symbols

☉ edible

(☉) edible after parboiling, drying, soaking, etc.

⊘ inedible, suspect, inferior

☒ poisonous

☒☒ very poisonous

☒(☉) poisonous, but edible after parboiling, drying, soaking, etc.

Foreword

Among the larger fungi the agarics are a uniform and clearly defined group which includes not only a large number of mycorrhizal fungi, but also many equally important species which live by breaking down dead matter. The agarics are therefore of great interest to fungi specialists and mushroom-pickers alike.

Since the agarics are particularly sensitive to environmental conditions (many are restricted to specific trees or grow on particular substrata) we have again laid great stress in this volume on carefully depicting their environment. The appearance of a fungus in relation to its setting – a fairy ring of Blewits scattered among autumn leaves, or a group of Parasol Mushrooms in a meadow glade – is often a useful guide to identification.

A great deal of intensive research has been concentrated on the agarics over the last twenty-five years. The results are gathered in Rolf Singer's book *The Agaricales in Modern Taxonomy*, a pioneer work first published in 1951 and revised twice since. We have gained a great deal from this work, as we have from several other modern works on agarics, whose authors include English, French and German researchers.

Bo Mossberg   *Sven Nilsson*   *Olle Persson*

The agarics

We described in the previous volume how men of earlier times were unable to understand the nature of fungi or to place them systematically in relation to plants and animals. The Florentine, P. A. Micheli, at the beginning of the eighteenth century, was the first to study fungi as a group. Linnaeus did little to advance the cause of mycology: his uncertainty was expressed in his extremely cautious pronouncements, and hesitation in placing them in the system he created for describing the world of organisms.

The large fleshy fungi were the first to be studied with any degree of care. The pioneers were the Dutchman C. H. Persoon and the Swede Elias Fries. Fries' splendid book *The Edible and Poisonous Fungi of Sweden* (1861) is the first illustrated Swedish work on edible fungi. Fries stimulated a general interest in them and a more intensive study of the agarics. Among the fungi experts whose works are invaluable, and still carefully studied, are the Italian G. Bresadola; the Frenchmen L. Quélet, H. Bourdot, A. Galzin, P. Konrad and A. Maublanc; the Englishmen M. Berkeley and M. Cooke; the German A. Ricken; the Finn P. A. Karsten, and the Dane J. E. Lange. One of the most important of modern critical works is R. Singer's *The Agaricales*. Sweden, through Elias Fries, led the way in the early study of fungi and his book promoted an interest in fungi in general, and in edible fungi in particular, which is reflected in a host of earlier popular books on the subject.

The Fly Agaric is well-known and is often used to depict fungi in general and the agarics in particular (even though most agarics have no ring or other remnants of a veil), and we have illustrated an example on the opposite page to represent them. Among those fungi which form fruiting hymenia, the agarics belong to the order of Agaricales. The boletes also belong to the Agaricales, while the clavarias, chanterelles, spine fungi and polypores belong to the order Aphyllophorales. Many amateur mycologists, especially those who pick mushrooms for food, are mainly interested in identifying quickly and easily the agarics and boletes. Certain genera, such as *Amanita*, of which there are only a few species in this country, are easily learned. On the other hand, the genus *Cortinarius* forms an enormous group with several hundred species, often closely resembling each other, which can only be mastered by a specialist.

To identify an agaric one must understand its structure. In the previous volume we described the typical development and structure, but by showing the structural variations it is easier to identify the genus. It is usually necessary to examine the whole fungus in order to make a safe identification. It may also be necessary to use a microscope. The following section deals briefly with the various parts of an agaric and the features which are important for purposes of identification. With the help of this summary, and by using the pictures and descriptions in this book, anyone should be able to identify a great many of our larger and more common fungi.

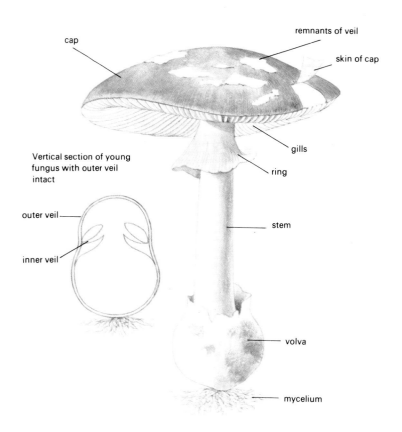

cap

remnants of veil

skin of cap

gills

Vertical section of young
fungus with outer veil
intact

ring

outer veil

inner veil

stem

volva

mycelium

The fruitbody flesh

The tissue in the cap and stem is usually called the flesh. It may be fibrous, tough, brittle or soft. With some genera, such as Fly Agarics, Parasols, Field Mushrooms and Ink Caps, the flesh of the cap and of the stem is not firmly joined, and the cap can be easily separated from the stem (see illustration p. 8). *Russula* and *Lactarius* flesh is brittle and easily breaks into pieces. *Lactarius* flesh contains a milky fluid. With other fungi the stem flesh splits into shreds.

The cap

The shape of the cap varies a great deal at different stages of development or under different climatic conditions even within the same species. Therefore, study both younger and older specimens. The cap skin may be slimy or sticky, as with *Gomphidius*, *Hygrophorus* and certain *Cortinarius* species. It is important to remember that the colour of the cap may vary with the age of the fungus and with the weather conditions. The skin of the cap can be easily peeled off in certain species.

7

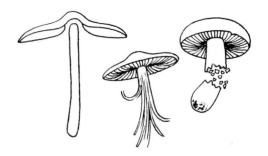

The consistency of the flesh may be decisive in identification: tough and fibrous or brittle and easily crumbled. With certain genera the cap can be easily pulled away from the stem without damage to either

Gill shapes

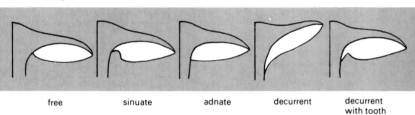

free   sinuate   adnate   decurrent   decurrent with tooth

Cap shapes

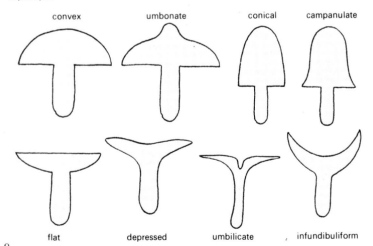

convex   umbonate   conical   campanulate

flat   depressed   umbilicate   infundibuliform

Examples of spore prints

The gills or lamellae

An important means of identification is the way the gills are attached. *Tricholoma*, for example, has sinuate gills. A number of *Hygrophorus, Clitocybe, Paxillus, Gomphidius* species and others have decurrent gills, while some have branched gills.

Remnants of the veil

Many agarics have remnants of the veil which surround the whole or parts of the young fruitbody. With Fly Agarics the veils remain as a ring and volva on the stem and as scales on the cap. *Cortinarius* species have a veil resembling a spider's web on the stem and at the edge of the cap. Species of *Amanita, Pholiota, Stropharia* and *Agaricus*, among others, have a ring on the stem.

The stem

The stem may grow either centrally or more or less to the side. It can sometimes be very short, or absent altogether. Its surface can have a very characteristic texture: fibrous, granular, smooth, etc. The stem base is sometimes swollen as in certain species of *Cortinarius, Macrolepiota* and *Clitocybe*.

The spores

The shape of the spores, often very characteristic of a genus, is not dealt with here because a detailed study demands a microscope. But when the spores – only some hundredths of a millimetre long – occur in large numbers, their colour is easily seen, which is an important factor in identifying genera. The spore powder may be white to yellow, brown, pink to red, or dark violet to black. By placing the fungus cap on a piece of white paper with the gills downwards, an attractive spore print is gradually produced (see illustration). White-spored fungi include *Amanita, Lepiota, Tricholoma, Clitocybe, Hygrophorus* and *Mycena*. *Entoloma, Volvariella* and *Clitopilus* have pink spores. *Pholiota, Hebeloma, Cortinarius* and *Inocybe* have brown spores, while in *Agaricus, Coprinus, Stropharia* and *Hypholoma* they are violet to black.

9

Scent and taste

Many fungi have a pronounced scent and taste, mainly in the gills. The ones that taste bitter or sharp are often poisonous, or at any rate inedible, although a number can be eaten after boiling. These are important characteristics but difficult to describe because no two people experience taste and smell in the same way. We try to describe a fungus's smell when it can be compared to a familiar substance. But if it smells of something less well-known, we cannot be so exact. An expert on the subject once described the smell of a certain species as 'like old lift-oil'. It was a very striking description, but how many people know what lift-oil smells like?

Edible fungi and mushroom-picking

Britain is something of a 'developing country' in gathering and using wild fungi, while some other countries are increasingly using fungi as a source of income and as a supplement to their food supplies. Finland has between 1,000 and 1,500 trained fungi 'consultants' and over 30,000 accredited pickers. The dairy industry, among others, takes care of the distribution and management. A limited number of fungi have been selected for the commercial trade. In the large wooded areas of Scandinavia it should be possible to harvest 200–300 kilos of fungi per hectare. In some areas bumper harvests of up to 3,000 kilos per hectare have been reported. In Britain only *Agaricus* is sold commercially, but wild mushroom collecting as a hobby is becoming increasingly popular and there are numerous organized expeditions.

However, there are many fungi varieties which are popular in eastern European countries but are not considered edible in Britain, such as *Lactarius turpis*, *L. rufus*, *L. trivialis* and *L. torminosus*. Their sharp and bitter substances disappear with boiling, salting and preserving. Millions of kilograms of woodland fungi are picked in Poland every year for preserving and for export. In some years nearly 100,000 pickers scour the woods for the preserved food industry. In Asia, USSR, and in Europe (Switzerland and France), dried fungi are important exports.

How much value do fungi actually have as food or nourishment? Their nutritive value has been debated for a long time, but in very general terms they can be compared with good vegetables. Many fresh and dried fungi, however, contain considerably more usable protein and carbohydrates than vegetables. They can easily absorb minerals, and their mineral content is perhaps the main reason why they are considered nutritious. In addition their fibrous substances are an aid to digestion. The table on the next page compares the nutritive substances of fungi with those of vegetables and milk.

	protein	carbo- hydrate	mineral substances	fat	water
commercial mushroom	3.6	3	1	0.2	90
Chanterelle	1.3	5	1.2	0.4	90
Penny Bun Bolete	5.4	5.2	1	0.4	87
Penny Bun Bolete (dried)	35.9	34.5	6.4	2.7	12
white cabbage	1.5	4.2	1.9	0.1	92
cucumber	–	1	0.4	–	98
milk	3.5	4.8	0.7	3.7	87
meat	21	0.5	1	5.5	72

The fungi's ability to convert carbohydrates and nitrogenous salts into valuable protein in the mycelium has been put to industrial uses, but in Britain it is still in the experimental stage. In several countries fungus protein is already a component of animal food and used as a supplement in human foods. Some day the use of wild fungi and the cultivation of fungus protein may partly revolutionize the food supply in many areas. In Britain, however, there are still reservations because the famines of the eighteenth century and rationing during the two world wars made fungi unpopular. It was regarded as second-class food, the stuff of emergencies.

We have not made any attempt to describe in detail mushroom-picking, the use of this or that edible fungus, nor included recipes, nor have we made any judgements on the most popular varieties. But some general advice to mushroom-pickers may be useful in this introduction to the agarics.

Mushroom-picking

Don't pick fungi that you cannot use. Let rare fungi stay where they are. When you want to identify unknown fungi they should be plucked with the base of the stem intact. Taste and smell the fungi, but only after learning to recognize the poisonous species. Avoid evil-smelling fungi and those with a sharp and bitter taste. Never pick mushrooms growing alongside busy roads, in newly-sprayed areas, on refuse dumps or industrial sites. Give well-known fungi a rough clean-up on the spot. Avoid unknown white fungi, especially if they are not fully grown. Pick fungi in dry weather, and use a basket or something similar rather than a polythene bag.

Never cook and eat fungi which you do not know with certainty to be harmless. If fungi are to be dried, this should be done very carefully and then they should be kept dry. Never try to keep cooked mushrooms, even in a refrigerator. At the first sign of any ill-effects after eating fungi, go to the nearest hospital at once.

Picking mushrooms is not only a more or less interesting preliminary to eating them. It can be a delightful hobby, or looked upon as good exercise and an opportunity to roam in woods and fields. But every mushroom-picker should show the same respect for fungi as for animals, flowers, bushes and trees. Many fungi are rare, and most species give more pleasure in the wood than in the pot.

Those who would like to increase their knowledge are urged to join a local society or the British Mycological Society, where they will be able to take part in organized forays and have the opportunity to consult experts.

Naming and classification

Classification is the branch of biology which deals with naming and classifying organisms and with the study of the relationships between them. A scientifically correct name is important for international communication and is essential for drawing up legislation, the administration of conservation organizations and in warning against the effects of dangerous organisms.

Fungi, like lichens, mosses and higher plants, have a Latin name consisting of two words. The first is the name of the genus, the second is the specific epithet. The two together form the name of the species. The name appearing after the specific epithet refers to the author of that species. This is often in a shortened form. For example, the Chanterelle is cited as *Cantharellus cibarius* Fr., where 'Fr.' refers to Elias Fries. Names sometimes have to be changed owing to revised opinions or new information. The name of the authority is then also changed. This is why after the names of many fungi, there may be a whole set of abbreviations.

Naming and classification are governed by the rules of nomenclature which have been ratified at successive International Botanical Congresses. The valid name for a fungus or plant is the oldest one which is correct according to the rules. Thus one does not go back in time indefinitely. Valid names of fungi date from certain nineteenth-century works of Fries and Persoon, and earlier names are therefore disregarded.

Regarding the assumption of relationships, organisms are placed in units, taxa (sing. taxon, cf. taxonomy) of different rank, each with a definite Latin ending. The most important are species, genus, family, order, class, division and kingdom. These taxa are put together into a clear system, reflecting what we believe to be the relationships of the units. In the case of fungi, numerous systems exist, but we detail here a simplified example of a much-used modern system, proposed in 1966 by the British mycologist C. G. Ainsworth. Certain classes and orders are omitted and to avoid making the system too large and too complicated some families have been left out.

Kingdom Fungi

Division *Myxomycota*, simple slime-moulds and related types; the vegetative stage in the life-cycle of a multi-celled body of protoplasm or dense grouping of amoebae; relationship with other fungi uncertain

Class *Myxomycetes*, slime-moulds, e.g. 'Flowers of Tan'

Division *Eumycota*, 'true' fungi

Subdivision *Zygomycotina*, thread fungi with immobile spores (i.e. not provided with flagella), e.g. pin-moulds

Subdivision *Ascomycotina*, ascomycetes, producing ascospores

Class *Hemiascomycetes*, including yeast fungi
Class *Pyrenomycetes*, flask fungi
Class *Discomycetes*, cup fungi

Order *Helotiales*, including earth-tongues
Order *Pezizales*, including morels
Order *Tuberales*, truffles

Class *Plectomycetes*, including mildews

Subdivision *Basidiomycotina*, basidiomycetes, producing basidiospores

Class *Hemibasidiomycetes*, basidial fungi, lacking a fruitbody.

Order *Uredinales*, rust fungi
Order *Ustilaginales*, smut fungi

Class *Hymenomycetes*, hymenial fungi

Order *Agaricales*, agarics, boleti
Order *Aphyllophorales*, polypores, etc.
Order *Tulasnellales*, jelly fungi, etc.

Class *Gasteromycetes*, stomach fungi

Subdivision *Deuteromycotina*, Fungi imperfecti; conidial stages which it has not been possible to connect with any of the above groups

Class *Hyphomycetes*, simple fungi with no fruitbody, commonly known as 'mould'
Class *Coelomycetes*, simple fungi with a fruitbody

Pleurotus ostreatus ⊖

(Jacq. ex Fr.) Quél. (from Greek *pleuron*: side and *otus*: ear, plus Latin *ostreatus*: like an oyster)

Oyster Mushroom

A number of rather tough fungi, with stems at the side of the cap and gills on the underside, can be found on wood, trunks and stumps. Some are mussel-like and even stemless. Many of these agarics were formerly placed in the family *Pleurotaceae*, but now are divided into various other families. One of the easiest to recognize is the Oyster Mushroom with a short stem joined to the side of the cap. The cap is brown to blue-grey (sometimes dark violet), shiny and 5–15 cm wide. This fairly common fungus in Britain often grows in tight clusters on either dead or living deciduous wood and on poplars and beeches, especially in parks and avenues. The Oyster Mushroom is now thought to be closely related to *Polyporus brumalis*, *P. squamosus* and other *Polyporus* species, and is placed in the family *Polyporaceae*. The cap is mussel-like, arched in varying degrees, with an inrolled margin. The gills are whitish, decurrent, and of varying length, often joining into a kind of net on the stem. The stem is white, 1–4 cm tall and up to 3 cm thick, with stiff hairs at the base. The scent is slight and the flavour mild, but the flesh is rather tough, so only young specimens should be eaten. Grows mainly from October to December, but in mild winters it can sometimes be found until April.

A strange fungus found on the branches and trunks of deciduous trees, especially beeches, is the Split Gill fungus *Schizophyllum commune* ⊗ Fr. (from Greek *schizo*: split and *phyllon*: leaf, plus Latin *communis*: common). Its caps are grey-white, hairy, leathery, furrowed towards the margin and are 2–4 cm in width. The fungus is unique because of its split gills, half of which roll themselves together when dry, and protect the hymenium, enabling it to survive drought. This fungus is almost restricted to southeast England, and is very rare in Scotland. It damages stored wood. Grows throughout the year when the weather is mild and damp.

Panellus serotinus ⊗

(Schrad. ex Fr.) Kühner (from Latin *panus*: a tumour, a name used by Pliny, plus Latin *serotinus*: late)

Somewhat smaller than the Oyster Mushroom, the cap is olive yellow to olive green and downy. The gills are yellow; the stem yellow-brown with small olive-coloured scales. Grows on deciduous trees from September to December.

Panellus mitis ⊗ (Pers. ex Fr.) Singer (from Latin *mitis*: mild) is a closely-related species with small mussel-like fruit-bodies and a stem attached to the margin of the cap. It grows on conifer twigs from October to February.

Schizophyllum commune

Pleurotus ostreatus

Panellus serotinus

Panellus mitis

Hygrophorus eburneus

(Bull. ex Fr.) Fr. (from Greek *hygros*: wet, damp, and *-phorus*: bearing, plus Latin *eburneus*: of ivory)

The wax agarics are named from their thick wax-like gills. Also, the basidia are longer than with other agarics. These criteria are subjective, but one soon learns to recognize a wax agaric. There are several groups of species: one containing more or less slimy species; one with dry ones, and yet another whose species have fragile watery flesh. The wax agarics illustrated are arranged according to colour: the whitish ones first; then the greyish ones and finally those with stronger colours. More abundant in wet weather, a number of species are found in meadows and pasture, others occur in woodland and are possibly mycorrhizal. One of the latter is the very slimy *H. eburneus* which is restricted to deciduous trees, including birch and oak. The ivory white cap, up to 8 cm wide, is at first convex, then broadens to umbonate. The gills are thick, wide-spaced and decurrent. The stem is yellow-white, speckled towards the top and 5–10 cm tall. This fungus often has a reddish tinge, seen most clearly on the gills. The smell is strong and disagreeable. It is not edible, and grows from August to October. A closely-related species *H. chrysaspis* Métrod is a less slimy, strong-smelling species and grows in beech woods.

Sections of Camarophyllus virgineus and C. pratensis. The gills of the latter are often connected by transverse ridges

Camarophyllus pratensis

(Pers. ex Fr.) Karsten (from Greek *kamara*: arch and *phyllon*: leaf, plus Latin *pratensis*: of the meadow)

Butter Mushroom

The Butter Mushroom belongs to the genus *Camarophyllus*: possessing a dry cap and decurrent gills. It reminds one of a Chanterelle, but the gills have sharp edges. It is common in grassy or mossy meadows or on dry hillsides, and is seldom attacked by insect larvae. It makes excellent eating. The 3–5 cm broad cap is first convex and somewhat conical, then becomes umbonate. The gills are widely spaced. The stem is firm and of even thickness, up to 8 cm tall. This fungus is first light orange, especially when damp, then pale leathery yellow. The scent is barely perceptible and the taste mild. The Butter Mushroom grows during September and October, and is common throughout northern Europe.

A closely-related species is *Camarophyllus virgineus* (Wulf. ex Fr.) Karsten (from Latin *virgineus*: virginal). It is at the most 4 cm wide and 7 cm tall and is never slimy. It has a mild taste and is edible, growing in grassy places during September and October.

16–17

Hygrophorus eburneus

Camarophyllus virgineus

Camarophyllus pratensis

Hygrophorus erubescens

Fr. (from Latin *erubescens*: blushing)

This pretty wax agaric, with its subtly varied pinkish colourings, is common and plentiful after long autumn rains in the shady coniferous woods of the Scottish Highlands. It may grow in large rings, and probably develops mycorrhizal associations with the conifers. Its bitter taste makes it unpalatable. It is, however, described as edible in a number of older books. Elias Fries wrote in 1861: 'The Blushing Wax Agaric comes high among our edible fungi.' More recent investigations have shown that there is a related but mild species which also grows in coniferous woods. It was first described in 1974, and also differs in that it does not turn yellow. Within this group of wax agarics (Section *Rubentes*) there are some other species. One of these, known in France as 'vinassier' (from French *vinasse*: weak wine, dregs of wine), grows under deciduous trees such as beech and oak, and is regarded in central Europe as a delicacy (see below).

Hygrophorus erubescens can vary a great deal. It may be almost white with a touch of pink, but is often whitish yellow, flecked with wine red. The entire cap usually has a red tinge, darker towards the centre. The yellowish tinge is clearest when the fungus is dry. The size can also vary a great deal. The cap is first convex, then flat with a downturned margin. It is sticky in damp weather, dry and shining in dry, fleshy, and up to 10 cm broad. The gills are fairly wide-spaced, somewhat decurrent, white to yellowish and usually speckled with red. The stem is white to yellowish, with pink granules on the upper part, firm, and 6–10 cm tall. The white flesh yellows somewhat when the fungus is cut. The scent is insignificant and the taste normally bitter, although it is sometimes mild. Grows from August to October.

The related species mentioned above, which grows in deciduous woods, is *Hygrophorus russula* ☉ (Fr.) Quél. (from Latin *russulus*: reddish). This is fleshier and a darker red, with more crowded sinuate gills and a shorter stem. A light brown species with a red tinge and a cap with a dark brown centre is *Hygrophorus discoideus* ☉ Fr. (from Greek *diskos*: plate and *eidos*: like). The cap is 4–6 cm broad and has decurrent yellowish gills. The stem is whitish, dotted towards the top and grows up to 9 cm high. The flesh is yellow-white to reddish, the scent is slight and the taste mild. The fungus, which is slimy in wet weather, is edible, and grows generally in coniferous woods on chalky ground, but is rare in Britain. It occurs in some places in Norway, mainly around the Oslo Fjord, which is well-known for its rich bedrock in chalk. Otherwise the distribution of this species has yet to be investigated.

Hygrophorus discoideus

Hygrophorus erubescens

Hygrophorus camarophyllus ⊖

(Alb. & Schw. ex Fr.) Dumée, Grandjean & Maire

This species has a pleasant flavour and makes an excellent dish. It is eaten also by animals and insects; an eminent Finnish mycologist, P. A. Karsten, wrote in 1879 that it was eaten voraciously by goats and snails.

The cap is sooty brown, with dark streaks running from the centre towards the rim. It grows up to 10 cm wide and there is often a distinct umbo in the middle. The gills are white-grey, thick, wide-spaced and decurrent. The stem is of even thickness, brown-grey, almost white where it joins the cap, and 8–12 cm tall. The white flesh is very brittle and the scent slight. The best time for picking it is during August to October when it thrives best in mossy, hilly coniferous woods.

Hygrophorus
camarophyllus
section

Hygrophorus hypothejus ⊖

Fr. (from Greek *hypotheios*: sulphur yellow on the underside)

Herald of the Winter

The night frosts are beginning when this late species first appears. It is easy to recognize by its slimy, olive brown cap and the decurrent yellow to orange gills. The cap is 3–6 cm wide, and the light yellow stem has a slimy 'stocking' which often ends in a barely perceptible ring. The fungus has a yellow flesh, and may grow up to 6 cm tall. The scent is slight and the mild flavour makes an excellent addition to a mixed mushroom dish. It grows plentifully in coniferous woods, often on mossy rocks, during October and November.

There are two other greyish, sticky to slimy species, also found in coniferous woods. One is *Hygrophorus olivaceo-albus* ⊖ Fr. (from Latin *olivaceoalbus*: olive and white). The cap is olive brown to olive grey, 3–6 cm broad, with a darker, slightly umbonate centre. The white gills are decurrent and the slimy stem is white above, flecked with brown below and grows up to 10 cm high. The scent is slight and the taste mild. This fungus is edible and grows in pine woods in Highland areas from August to October. *Hygrophorus agathosmus* ⊖ Fr. (from Latin *agathosmus*: pleasant-smelling) has a light grey cap which is somewhat convex, sticky, and 4–7 cm broad. The white gills are widely spaced and decurrent and the stem is white to greyish, with white speckles towards the top, 8–10 cm high. The strong scent suggests bitter almond oil or almond soap. The taste resembles the smell, rendering it unpalatable. It grows generally between August and November, especially among moss in pine woods.

Hygrophorus
olivaceoalbus

Hygrophorus
hypothejus
section

Hygrophorus camarophyllus

Hygrophorus hypothejus

Hygrophorus agathosmus

Hygrocybe punicea ☉

(Fr.) Karsten (from Greek *hygros*: wet, damp and *kybe*: head [cap], plus Latin *puniceus*: purple-coloured)

Most *Hygrocybe* species are brilliant red or yellow and grow in pastures, groves and meadows. As far as we know, none form mycorrhizal associations. The genus consists mainly of small to medium-sized species. The largest and most impressive is the common *Hygrocybe punicea* which grows in grassy places in September. If the weather is mild it may be found late into the autumn. The cap is at first bell-shaped and blood red, and then spreads out to become more yellowish, glossy, and in wet weather somewhat sticky. It is 5–10 cm wide. The gills are thick, light yellow-red, and widely spaced, while the stem is yellow-red, whitish and streaky towards the base. The scent is slight, the taste mild, and this fungus is good to eat and much sought after, but should not be confused with *Hygrocybe conica* or *H. nigrescens*.

A closely related but smaller species, also scarlet, is the Scarlet Hood *Hygrocybe coccinea* ☉ (Schaeff. ex Fr.) Kummer (from Latin *coccineus*: deep red). This is no wider than 5 cm and has a yellow to scarlet stem with no streaks, and yellow-red gills. The gill edges often have a bluish tinge. This fungus is edible, and grows from August to October in the same type of soil as *H. punicea*.

Hygrocybe coccinea

Hygrocybe conica ⊗

(Scop. ex Fr.) Kummer (from Latin *conicus*: conical)

Conical Slimy Cap

A fairly slender yellow-red species, at the most 5 cm in width and 10 cm tall. It blackens when old or when touched. The cap is pointed at first, then broadens, retaining a pointed umbo. The gills are at first almost white, but later are greyish olive to orange, and finally blacken. The yellow stem is often somewhat twisted. The scent is slight and the taste mild, but this fungus should not be eaten. There are in fact several closely-related, pointed, orange-coloured species which are suspected of being poisonous. These grow from July to October on grassy land throughout northern Europe.

Another beautifully-coloured species is *Hygrocybe chlorophana* ☉ (Fr.) Karsten (from Latin *chlorophanus*: greeny-yellowish). The lemon yellow cap has whitish streaks running from the centre to the margin and is 2–4 cm broad. The yellow stem is irregular and often flattened, 4–6 cm tall. This fungus is slimy in damp weather but it has a slight scent, a mild taste, is edible and grows where grass has been cropped short by animals. It occurs from August to October.

22–23

*Hygrocybe
punicea*

*Hygrocybe
chlorophana*

*Hygrocybe
...onica*

Calocybe gambosa ⊖

(Fr.) Donk (from Greek *kalos*: pretty and *kybe*: head (cap), plus Latin *gambosus*: courtly)

St George's Mushroom

The shiny white fairy rings of St George's Mushroom stand out against the luxuriant green after warm summer rain. It thrives on calcareous soils, and is common in northern Europe. The mushroom is now placed in the genus *Calocybe*, of which there are nine species in Britain. In Elias Fries' classic grouping all the agarics with whitish spores and sinuate gills were classified under the genus *Tricholoma* (from Greek *trichos*: hair and *loma*: fringe). Today these fungi are divided into numerous genera. The division mainly depends on microscopic features, but colour and scent can help in field identification. Thus we deal first with white Tricholomas, then yellowish ones, and then brown ones.

*Calocybe
gambosa*

The St George's Mushroom cap is first umbonate and then convex. It is white (buff-coloured in damp weather), irregularly wavy, fleshy and 5–10 cm broad. The dense gills are sinuate where they join the stem, which is normally stout, white to yellow-white, 4–8 cm tall. The strong scent suggests damp meal but the full-bodied good flavour is particularly excellent in soup. Found from the end of April to June, but it may also appear in the autumn.

St George's Mushroom is sometimes confused with the white to yellowish *Melanoleuca strictipes* ⊖ (Karst.) J. Schaeff. (from Greek *melas*: black and *leukos*: white, plus Latin *strictipes*: straight-stemmed). The genus *Melanoleuca* is characterized among other things by a cap, with a distinct umbo, which grows up to 10 cm wide, and by a stiff cylindrical stem up to 12 cm tall. Its scent suggests yeast. It is mildly poisonous and grows singly or in small groups in pastures from June to September.

Tricholoma album ⊗

(Schaeff. ex Fr.) Kummer (from Latin *albus*: white)

This dry, white to yellowish Tricholoma, found throughout northern Europe, is restricted to birch trees. It belongs to the genus *Tricholoma* in the strict sense and is easily recognized by its nauseous smell. The cap is first convex with an in-turned margin and then broadens out and grows up to 12 cm wide. The gills are crowded and the stem is stout, 5–8 cm tall. It has a bitter taste, is not edible, and grows from August to October, often in large rings.

*Melanoleuca
strictipes*

24–25

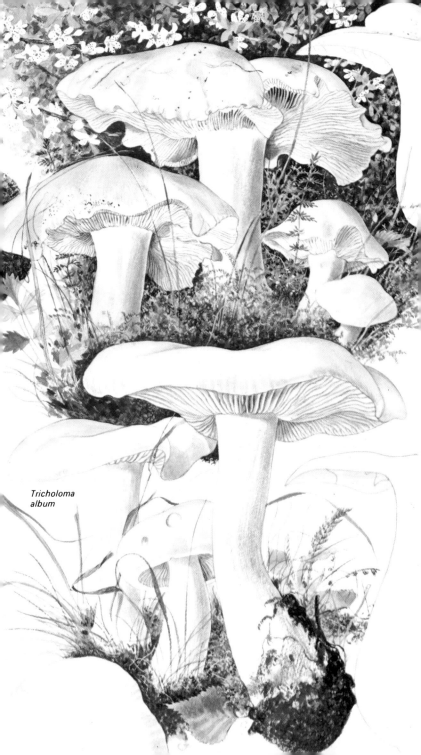

*Tricholoma
album*

Tricholoma flavovirens ☉

(Pers. ex Fr.) Lundell (from Latin *flavovirens*: yellow-green)

A typical Tricholoma with its smooth spores and white spore print, it is fairly firm and sturdy and conspicuous because of its sulphur yellow colour. It is highly prized for eating, especially in southern Europe but it is uncommon in Britain, occurring in sparse, open coniferous woods. This fungus often develops below the surface and the cap, sticky in damp weather, becomes covered with pine needles and debris. The delicately scaly or radially fibrous cap is 5–10 cm broad, a brilliant sulphur yellow tinged with brown. The gills are a beautiful sulphur to orange-yellow and the stem is also yellow, up to 6 cm tall. The strong scent and the flavour of fresh damp meal help make a delicious mixed mushroom dish.

Tricholoma flavovirens is sometimes confused with *Tricholoma aestuans* ☒ (Fr.) Gill. (from Latin *aestuans*: burning, glowing). This is a smaller and slimmer fungus with a very bitter taste. Another similar fungus is the Sulphurous Tricholoma *Tricholoma sulphureum* ☒ (Bull. ex Fr.) Kummer (from Latin *sulphur*), a fairly small species found in deciduous woods, with a smooth sulphur-coloured cap and widely-spaced gills. The smell resembles coal gas and the taste is bitter. Both these species are poisonous.

Tricholomopsis rutilans ☉

(Schaeff. ex Fr.) Singer (from Latin *tricholomopsis*: like Tricholoma and *rutilans*: reddish)

In the genus *Tricholomopsis* we place a number of species which grow saprophytically on dead wood. There are three species in Britain: the common *T. rutilans*, *T. platyphylla* ☉ (Pers. ex Fr.) Sing., and the less common *T. decora* ☉ (Fr.) Sing.

Tricholomopsis rutilans occurs on decaying conifer stumps and on roots, mainly pines. Generally it grows in clusters, appearing in early summer. It is common in Britain and widely distributed throughout northern Europe. It is a fairly large species with a cap up to 15 cm broad. The cap has a yellowish background, densely covered with reddish scales. The yellow stem is tall and stout and the yellow flesh tastes bitter. It is edible after boiling but cannot be recommended.

Tricholoma sulphureum

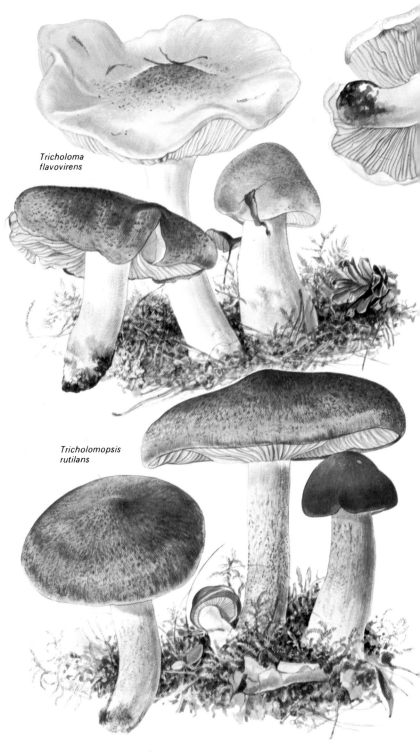

*Tricholoma
flavovirens*

*Tricholomopsis
rutilans*

Tricholoma albobrunneum ☠

(Pers. ex Fr.) Kummer (from Latin *albobrunneus*: white-brown)

This brown sturdy Tricholoma can be seen in coniferous woods as early as mid-August. The cap, with its radial streaks, is sticky in damp weather. This common fungus occurs throughout the autumn into November mainly in pine woods on a sandy soil, but also in other types of coniferous and mixed woodland.

Tricholoma albobrunneum grows to a height of 10–15 cm. The cap, brown to dark brown with darker shades at the centre, is about 10 cm wide, convex and at first has an inrolled margin. The white gills are later flecked with brown and the cylindric stem is white above, brownish below. The flesh smells slightly of meal but the mild taste leaves a bitter after-taste. This is a poisonous species, and severe cases of poisoning have been reported. It is wise to avoid eating any brown Tricholomas because there are no good edible species among them.

Less common than *T. albobrunneum* is *Tricholoma pessundatum* ☠ (Fr.) Quél. (from Latin *pessundatus*: sunk, destroyed). In some years it can be found in great numbers. The cap of this species has round drop-like markings and no radial streaks. It is also poisonous.

Tricholoma albobrunneum can be confused with *T. fulvum* ⊖ (DC ex Fr.) Sacc. (from Latin *fulvus*: orange, yellow-brown). This is similarly coloured, and also has a cap which is sticky when damp. The cap, however, tends to be pale yellow towards the margin. The yellow-white gills later become rust coloured with brown flecks. It grows in deciduous woodland and is common throughout northern Europe. There are two other quite common species of brown Tricholoma with more or less scaly caps. *Tricholoma vaccinum* ⊖ (Schaeff. ex Fr.) Kummer (from Latin *vaccinus*: cow) normally grows in coniferous woods from August to November. The brown to red-brown cap is 5–10 cm broad and is at first a rounded bell-shape with an inrolled margin. Later it broadens and has a hairy, beard-like margin. Its dingy white stem is red-brown and fibrous towards the base. The flesh, which reddens when touched, tastes bitter. *Tricholoma imbricatum* ⊖ (Fr. ex Fr.) Kummer (from Latin *imbricatus*: patterned like roof-tiles) is rather like *T. vaccinum* but larger. The darker brown cap has small dark scales. The gills are first white, then flecked with brown. The stem is pale above and darker towards the base. It grows in pine woods on a sandy soil.

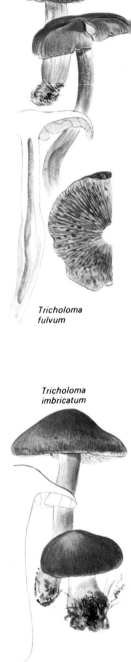

Tricholoma fulvum

Tricholoma imbricatum

*Tricholoma
albobrunneum*

*Tricholoma
vaccinum*

Tricholoma saponaceum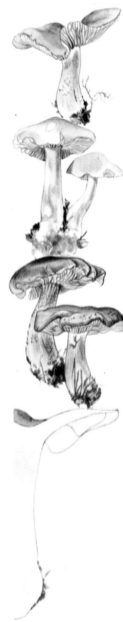

(Fr. ex Fr.) Kummer (from Latin *saponaceum*: soapy)

Soap-scented Tricholoma

The Soap-scented Tricholoma, named from the fairly strong smell of soap or newly-scrubbed floors, is very variable in colour and manner of growth. It is normally found in coniferous and mixed woods, but it crops up here and there in entirely deciduous woods and on more open land, especially near the coast. In southern and central Europe it often grows under oaks. It is quite common throughout Britain from the end of July to October. The cap, 5–10 cm broad, is convex and bell-shaped, later broadens irregularly and often has an umbo. The dull colour varies from yellow-white to brown, nearly always with tinges of grey or green. The pale yellow or yellow-green gills have a tinge of pink. The stem is dirty white to grey-green, up to 8 cm tall. The Soap-scented Tricholoma often has patches of pale red, and the flesh develops red patches when pressed. The taste is mild, but this fungus is slightly poisonous.

Tricholoma portentosum

(Fr.) Quél. (from Latin *portentosus*: marvellous, monstrous)

A sturdy, fairly large species, sometimes confused with the Soap-scented Tricholoma. It differs, however, in having radial lines or streaks on the cap, which becomes sticky when damp. It grows fairly generally throughout the autumn on the ground in coniferous woods, mainly pine. Occasionally, however, it appears in mixed woods.

It is similar to *Tricholoma sejunctum* ⊖ (Sow. ex Fr.) Quél. (from Latin *sejunctus*: separated) but has a more yellowish cap with brownish streaks. The gills are white to pale yellow. The pale margin of the cap is an important feature for recognition. This fungus is not uncommon in mixed woods.

Another grey-coloured Tricholoma is *Tricholoma virgatum* ⊖ (Fr. ex Fr.) Kummer (from Latin *virgatus*: streaky) with a shining, silver grey cap with a pointed umbo and dark, innate, radial striations. The flesh smells of damp earth, and has a very sharp and bitter taste, distinguishing it from other grey-coloured species. *Tricholoma virgatum* is less common than *Tricholoma portentosum* and grows in both coniferous and deciduous woods, especially beech.

There is a similar grey, but scaly, common Tricholoma which one sees at roadsides, in well-trodden pastures and in parks. This is *Tricholoma terreum* ⊖ (Schaeff. ex Fr.) Kummer (from Latin *terra*: earth).

Variations of shape and colour in Tricholoma saponaceum, *with a vertical section*

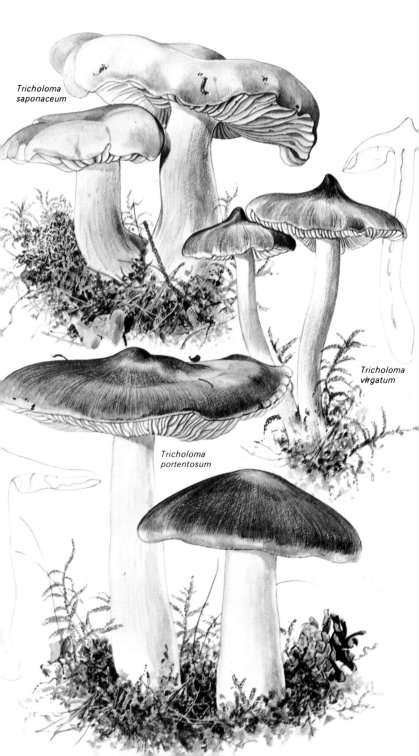

*Tricholoma
saponaceum*

*Tricholoma
virgatum*

*Tricholoma
portentosum*

Lepista nuda ⊖

(Bull. ex Fr.) Cooke (from Greek *lepista*: bowl and Latin *nudus*: naked)

Wood Blewit

The Wood Blewit is undoubtedly one of the best-known species. Its striking appearance comes from the bluish colour, and many people regard it as a delicacy. Found in colonies, often in fairy rings, especially on land that has been cultivated, it grows on decaying leaves, piles of straw and compost and sometimes on the carpet of fallen needles under pine trees. It is widely distributed throughout northern Europe, and is sometimes seen very late in the autumn if the weather is mild and damp. The genus *Lepista* is characterized by spiny spores and a pinkish spore print.

The Wood Blewit is fairly stout, and blue-lilac to blue-grey in colour. Older specimens are often brownish. The cap is first convex, then it expands and develops an uneven margin. It is 5–15 cm broad. The sinuate gills are blue-violet, and easily separated from the cap. The stem is of even thickness, 5–10 cm tall and blue-violet, becoming whitish towards the top. The slightly aromatic scent sometimes suggests warm rubber. The taste is mild and the fungus is edible, although some people consider it overrated. It can be confused with the violet *Cortinarius* species, particularly *Cortinarius traganus*, and *C. camphoratus* (see p. 90). However, these have conspicuous brown-coloured remnants of their 'spider's web' veil on the stem and cap margin. Moreover, their taste is sharp and the spore print is brown.

Lepista saeva ⊖

(Fr.) P. D. Orton (from Latin *saevus*: furious)

Colour variations in
Lepista nuda

Blewit

This pretty, two-coloured fungus, with a grey-brown cap and mauve to blue-lilac stem, is closely related to the Wood Blewit. It is found in rich meadows, deciduous woods, parks and at the edges of woods. Both may grow in rings. It is common and widespread in Britain in the late autumn but it is thought to have decreased in numbers because of changes in land use leading to a reduction of suitable habitats.

The cap is first convex, then expands to a breadth of 5–15 cm. The crowded gills are greyish and the stem, 5–10 cm tall, is a particularly brilliant lilac colour when the fungus has grown in deep grass. It smells of meal, has a mild taste and is good to eat.

Lepista nuda

Lepista saeva

Lyophyllum connatum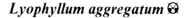

(Schum. ex Fr.) Singer (from Greek *lyo*: I release, I set free and *phyllon*: leaf, plus Latin *connatus*: growing together)

Related to Tricholoma is a group of fungi with firmer, more elastic flesh and somewhat decurrent gills. These also have carminophilous basidia, i.e. basidia in which reddish granules can be seen when stained by aceto-carmine. These used to be placed in the genus *Tricholoma*, or other genera such as *Clitocybe* and *Pleurotus*. Some grow in clusters like the species described here.

One occasionally sees clusters, long rows or single, almost pure white, *Lyophyllum connatum* along overgrown gravel roads and old cart tracks. Found mainly on land that has been cultivated, as well as at roadsides and in groves and parks, the cap is pure white in dry weather and greyish when damp. It is up to 10 cm broad, irregularly convex and has a wavy margin. The white gills are somewhat decurrent in older specimens. The stem is white to grey-white, thicker towards the base, 5–8 cm tall. The fairly strong scent is somewhat suggestive of alkali. This fungus is edible when young, but should be boiled first. Some people, in fact, become mildly ill if they eat it without boiling. It grows from August to October.

Lyophyllum aggregatum

sensu lato (from Latin *aggregatus*: crowded together)

Lyophyllum aggregatum is a collective name for a group of closely related grey to grey-brown fungi which have not yet been fully investigated, and which were previously described under a number of different names. They often occur on fertile grassy land that has been cultivated, but also in woods. The yellow-grey to brown cap is usually 10–15 cm broad. The stem and the gills are white in some species, greyish in others. The form of the gills also varies: they may be adnate or subdecurrent. The tough elastic flesh is edible, although some specimens have a somewhat burning taste. Because this complex of species has been only sketchily researched, it is wisest to avoid eating them.

One of the better-known fungi now placed in the genus *Lyophyllum* is *Lyophyllum ulmarium* (Bull. ex Fr.) Kühner (from Latin *ulmarius*: elm) previously called *Pleurotus ulmarius*. In these species several fruitbodies often grow together from a common base.

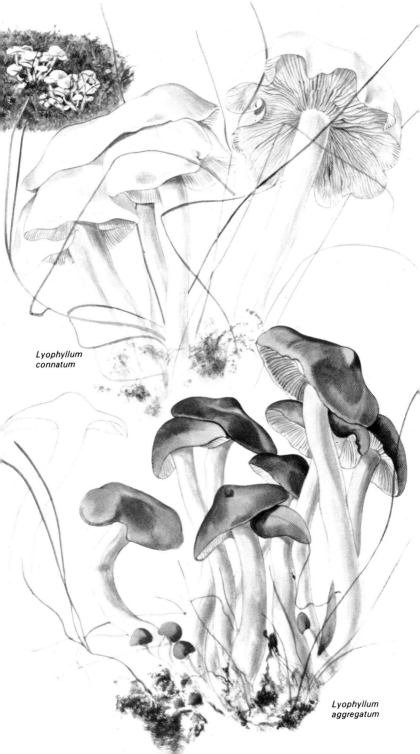

*Lyophyllum
connatum*

*Lyophyllum
aggregatum*

Armillaria mellea ☉

(Vahl ex Fr.) Kummer (from Latin *armilla*: ring, and Greek *meli*: honey)

Honey Fungus

It is an eerie experience to walk through a wood where the Honey Fungus has grown unchecked – thousands of fruitbodies cover stumps, branches and roots. In our woods there are a large number of small and middle-sized fungi which grow generally on stumps and branches. Some are parasites, others are saprophytes. Of those which have a ring the best-known is the Honey Fungus.

The Honey Fungus is one of the most destructive of fungi in forested land, causing losses amounting to millions of pounds in Europe alone. The worst ravages are seen in deciduous woodland, but it will also attack fir trees, and even such plants as phlox and rhubarb. When young the fungus is dirty yellow to honey yellow with a touch of brown; when older it is almost wholly brown. The scaly cap is first globose, then expands to a breadth of 10 cm. The young gills are dirty white to yellowish. The stem is 10–15 cm tall and sometimes curved. The spores, formed in great quantity, often cover the fungus and its surroundings in a whitish powder, sometimes in a layer that looks like snow. This fungus has an unusual mycelium with repeatedly branching threads called rhizomorphs (from Greek *rhiza*: root and *morphe*: shape). They grow to several metres and may be very broad and thick. Their outer parts form a black crust. If these threads force their way under the bark of branches and stumps, they often become thin and ribbon-like. Under the bark of dead trees one nearly always finds a network of these 'boot laces' which, curiously, shine (phosphoresce) in the dark. This fungus, common throughout Europe, has a mild scent and taste. Young caps with white gills are edible.

The beetle Anisotoma humeralis may darken the gills of Armillaria mellea

rhizomorphs (mycelium)

Laccaria laccata ☉

sensu lato (from Latin *laccatus*: shining, varnished)

The Deceiver

The Deceiver has a cartilage-like stem, and wide-spaced gills growing particularly far apart at the stem. The flesh is thin and watery. The light red to salmon-coloured types can be divided into several species, each with a particular mode of growth. One species which is plainly distinct from these is the light to dark violet Amethyst Deceiver *Laccaria amethystea* ☉ (Bull. ex Mérat) Murr. (from Latin *amethysteus*: amethyst-coloured). All these species are edible but tasteless. They are extremely common in late summer and autumn in woods, meadows and pastures.

Vertical sections of Laccaria laccata *and* L. amethystea

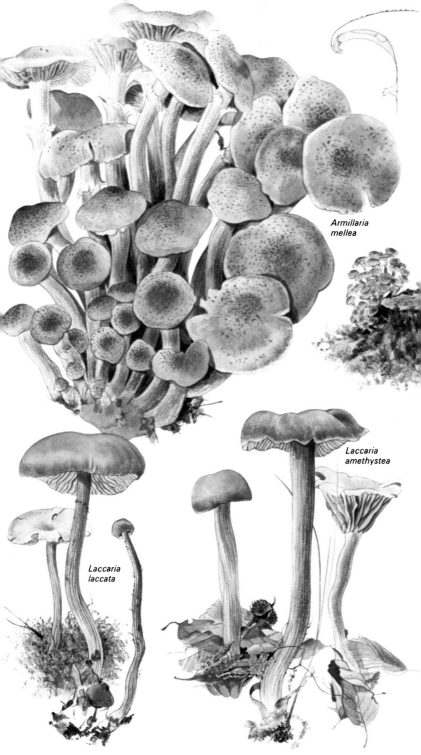

*Armillaria
mellea*

*Laccaria
amethystea*

*Laccaria
laccata*

Leucopaxillus giganteus ⊙ (✷ allergy)

(Fr.) Singer (from Greek *leukos*: white, and Latin *paxillus*: plug, plus Latin *giganteus*: gigantic)

Giant Clitocybe

Clitocybe species are closely related to *Tricholoma*, but recently certain species have been removed from the old genus *Clitocybe* (from Greek *clitos*: bent and *kybe*: hat) and transferred to other genera. They are characterized by a more or less funnel shape, emphasized by gills which run down the stem; in many species the cap itself is depressed. The spore print is white. The genus embraces some hundreds of species, of which there are about sixty in Britain. Many grow on open grassland and cultivated soil.

The largest species is *Leucopaxillus giganteus*, whose cap can grow to a width of about 30 cm. The fully-grown cap has a funnel shape with a thin margin. The colour is white to chamois. The gills are white at first, and later become leather-coloured. The stem is 5–10 cm tall. The firm flesh is fibrous with a slight scent of almond blossom, or, with older specimens, of milk. The strong taste is almost aromatic, and young specimens are good to eat. As with a number of other fungi, even good and edible *Clitocybe* species may cause mild illness in some people if not boiled before eating, and some people may be seriously allergic to them. *Leucopaxillus giganteus* is found from July to September in fertile soil in pastures, parks, the glades of deciduous woods and at the edge of woodland. It often forms enormous fairy rings.

Clitocybe geotropa ⊙

(Bull. ex Fr.) Quél. (from Greek *ge*: earth and *tropos*: turn)

This creamy white to chamois-coloured fungus is another large species. It has a well-developed mycelium lying deep in the soil, often grows in semicircles and is easily recognized. The cap of a young specimen is small in relation to the tall stem and has an inrolled margin. The cap, which has a pronounced umbo, grows up to 15 cm broad. The stem, 10–20 cm tall, widens somewhat towards the base. The aromatic scent is markedly spicy, not unlike that of bitter almonds. It grows fairly commonly during the autumn in copses, deciduous meadows, woodland clearings and at the edge of woods. Although this fungus is good to eat when young, there are several other more or less white species, many of which are poisonous or inedible. They are easily confused with *Clitopilus prunulus* (see p. 98) but have white spore prints instead of the pink spores of that species.

*Leucopaxillus
giganteus*

*Clitocybe
geotropa*

Clitocybe infundibuliformis ⊖

(Schaef. ex Weinm.) Quél. (from Latin *infundibuliformis*: funnel-shaped)

Common Funnel Cap

The genus *Clitocybe* includes a number of more or less brown species which are often hard to separate. They are fairly common woodland fungi. *Clitocybe infundibuliformis* sometimes resembles small specimens of *C. geotropa*. The cap is first convex, then expands and finally becomes deeply funnel-shaped. It is 5–10 cm broad, often with a wavy edge. The white gills are widely spaced and run far down the stem, which is 4–8 cm tall, tough and tapers upwards. The cap and stem are yellow-brown to light brown-red. This fungus has roughly the same scent as *C. geotropa* but a milder taste. Like several related species, it has thin flesh and is not particularly good to eat. Common throughout the country, it grows profusely in deciduous woods and in mixed woods from July to September, often in large groups or rings.

Clitocybe flaccida ⊗

(Sow. ex Fr.) Kummer (from Latin *flaccida*: flaccid)

Very similar to the preceding species, but has a stronger yellow-brown to red-brown colouring and often grows around heaps of decaying leaves and twigs. It is also found in woods. A late autumn fungus which may fruit far into November, it often grows in rings or groups like other *Clitocybe* species. The cap, 5–10 cm broad, sometimes has darker patches. The gills are flecked with brown and the spore print is brownish white. The stem, 3–6 cm tall and lighter in colour, is sometimes curved and has a felt-like hairy texture at the base. The smell in older specimens is somewhat earthy. This fungus has a mild taste, but cannot really be recommended for eating.

    The Blue-green Clitocybe *Clitocybe odora* ⊗ (Bull. ex Fr.) Kummer (from Latin *odorus*: pleasant-smelling) has a strong aniseed scent and is conspicuous for the pretty grey-green to white-green colour. The gills are not markedly decurrent and the relatively slender stem is often curved at the base. The species is certainly edible, but it can be confused with poisonous ones and should therefore be avoided. It is fairly common in litter of deciduous woodland.

Section and base of stem of Clitocybe flaccida *and section of* C. odora

Clitocybe infundibuliformis

Clitocybe odora

Clitocybe flaccida

Clitocybe nebularis (⊖)

(Batsch ex Fr.) Kummer (from Latin *nebula*: mist, cloud)

Clouded Agaric

A fairly large and sturdy agaric which is found in leaf litter in both deciduous and coniferous woodland, but occurs also in parks, gardens, copses and at the edge of woods. It often grows in large rings or long rows, sometimes 10 metres long. It thrives on rich soil, and occurs commonly in autumn throughout the British Isles.

This fungus is light grey to dark grey in colour, but more or less white specimens can also occur. The cap is first convex, then flat, 5–15 cm broad. The upper surface is often powdered with spores and has a characteristic whitish bloom. The gills are crowded and slightly decurrent. The stem is lighter in colour than the cap, sturdy, somewhat thicker towards the base and may grow up to 12 cm tall. The smell of the flesh is faint but sweetish, and the flavour, when cooked, is highly aromatic. 'I have eaten it fried with bacon', writes Elias Fries of this controversial fungus, and recommends drying it for winter use. Obviously it suited him well. This is not true of everyone who eats it, however. Many people do regard it as a good edible fungus, but it may cause illness in some. The very poisonous *Entoloma sinuatum* may also be found nearby. This, however, has pinkish, notched, more widely-spaced gills, and a pink spore print (see p. 98).

Clitocybe clavipes ⊖

(Pers. ex Fr.) Kummer (from Latin *clava*: club and *pes*: foot)

This species is smaller than the Clouded Agaric, and darker in colour. It grows in both deciduous and coniferous woods in early autumn, and is common throughout Britain and northern Europe. The cap is 3–8 cm broad, sooty brown to grey-brown. The gills are widely spaced and deeply decurrent. The stem is paler in colour, has a club-shaped base, and grows up to about 8 cm tall. The flesh is somewhat watery and has a slight sweet scent which suggests bitter almond. The taste is mild and the fungus is edible.

A species which was previously placed in the genus *Clitocybe* (see p. 44) is *Cantharellula cyathiformis* ⊖ (Bull. ex Fr.) Singer (from diminutive of *Cantharellus*, plus *kyathos*: a cup, and Latin *forma*: shape). This is a medium-sized fungus (5–10 cm tall) with a deeply funnel-shaped cap which is dark coffee brown when damp. The straight stem is the same colour. This fungus is edible. It usually fruits in late autumn, and occurs in woods and parks, in grass and among piles of leaves and twigs.

Cantharellula cyathiformis

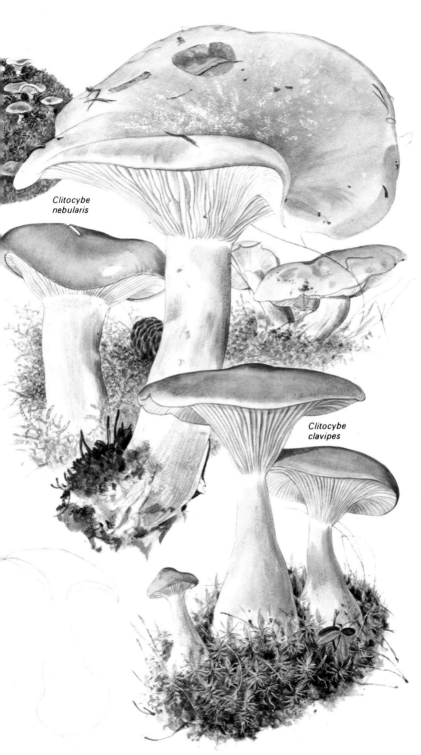

*Clitocybe
nebularis*

*Clitocybe
clavipes*

Omphalina ericetorum ⊗

(Fr. ex Fr.) M. Lange (from Greek *omphalos*: navel, and Latin *ericetorum*: heath)

Umbrella Navel Cap

The fungi commonly called Navel Caps are closely related to *Clitocybe* species, and their main features are a more or less funnel-like shape, a wax-like or cartilaginous consistency and a hollow stem. The cap, pale yellow to brownish olive with darker furrows, is 1–2 cm in breadth with an umbilicate depression. The broad, widely spaced, decurrent gills are pale yellow. The short stem is the same colour as the cap. This fungus often grows in mossy places, on decaying wood and on peat.

The rare species *Xeromphalina campanella* ⊗ (Batsch. ex Fr.) Kühn & Maire (from Latin *campus*: field) is rust yellow to yellow-red and somewhat smaller. Its stem narrows at the bottom and it grows on the decaying stumps and branches of conifers.

Mycena galericulata ⊗

(Scop. ex Fr.) S. F. Gray (from Greek *mykes*: fungus, and Latin *galericulum*: little hood)

Bonnet Mycena

The genus *Mycena* is related to the genus *Omphalina* and similar in appearance to the genus *Marasmius* (see p. 48). *Mycenas*, however, are not so tough and long-lived. As the name suggests, the light grey to brown-grey cap has a bell-like or bonnet-like shape 2–5 cm broad, with a markedly 'pleated' appearance. The thin gills are faintly pink. The stem is 3–8 cm tall, rigid, shiny grey and bristly at the base. This fungus grows on rotting wood from spring to late autumn.

Another of the larger species is *Mycena pura* ⊗ (Pers. ex Fr.) Kummer (from Latin *purus*: pure, unmixed). This entire fungus is rose to lilac-coloured. It has a cap 2–8 cm across, widely-spaced gills and a tall, hollow stem 4–8 cm high which is white and downy at the base. It generally grows on decaying matter in both deciduous and coniferous woods.

Conspicuous among the small species is the yellow-stemmed *Mycena epipterygia* ⊗ (Scop. ex Fr.) S. F. Gray (from Greek *epi*: on, and *pteris*: fern) with a shiny sticky cap whose skin is easily pulled off.

One of the prettiest species is the tiny and delicate *Mycena rosella* ⊗ (Fr.) Kummer (from Latin *rosellus*: slightly rose-coloured). It grows in damp, mossy, coniferous woods.

In mild, damp weather in late autumn, or during a mild winter, you may find a small species growing on the bark of deciduous trees. This is *Mycena corticola* ⊗ (Pers. ex Fr.) Quél. (from Latin *corticola*: bark), a delicate fungus about one centimetre high, which is grey or greyish red.

In late autumn Mycena corticola *often grows on the bark of limes*

Mycena corticola

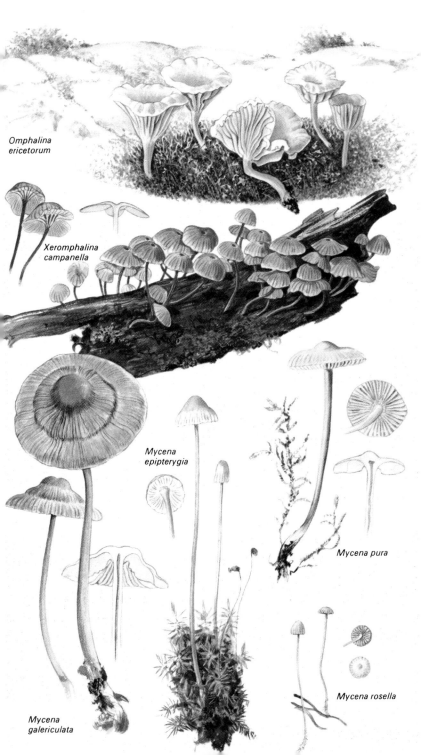

*Omphalina
ericetorum*

*Xeromphalina
campanella*

*Mycena
epipterygia*

Mycena pura

*Mycena
galericulata*

Mycena rosella

Flammulina velutipes ⊖

(Curt. ex Fr.) Karsten (from Latin *flammula*: little flame, *vellus*: pelt and *pes*: foot)

Velvet Shank

Among the agarics there are some groups of fairly small fungi with tough, almost elastic flesh. The Tough Shanks are one such group, their main feature being a tough, elastic and hollow stem. Many of them are small and easily missed. They were previously placed in the genus *Collybia*, but many of them have now been placed in other genera, such as the related *Marasmius* (see p. 48).

One of the more recognizable species is the Velvet Shank which appears throughout the winter if the weather is mild. The slimy and streaky cap is 3–6 cm across, honey-coloured to yellow-brown with touches of red. The gills are a dirty brown, the spores white. The upper part of the stem is off-yellow while the lower part is brown with a dark, downy, velvety texture. The scent and taste are mild, and the fungus is edible. It is found in rows or clusters at the bottom of trees, on stumps or on roots, especially elm or aspen. The species is very common in Britain.

Oudemansiella radicata ⊖

(Relhan ex Fr.) Singer (from *Oudemans*, the name of a Dutch botanist, and Latin *radix*: root)

Rooting Shank

The common name describes the main feature of this fairly large agaric: the long root-like stem. The wrinkled cap is 5–10 cm wide, browny grey to brown-yellow or olive brown, and sticky in wet weather. The dirty grey-yellow stem is slender and tough. The scent and taste are mild, and the fungus is edible, although it is far from tasty. Found in summer and autumn on the stumps of deciduous trees, it is fairly common throughout northern Europe.

Close in colour is the Greasy Tough Shank *Collybia butyracea* ⊖ (Bull. ex Fr.) Quél. (from Latin *butyrium*: butter) with a red-brown cap with a darker centre. The stem is brown. It grows commonly both in deciduous and coniferous woods, often in clusters or rings, and is widespread throughout the country.

Collybia acervata

Collybia dryophila

Another species is the very common Russet Shank *Collybia dryophila* ⊖ (Bull. ex Fr.) Kummer (from Greek *dryos*: oak and *philos*: loving). In damp weather it is yellow-brown to red-brown; in dry weather pale, almost yellow-white. It grows throughout the summer and autumn, especially in deciduous woods. Less common is the closely-related *Collybia acervata* ⊖ (Fr.) Kummer (from Latin *acervatus*: clustered). The cap is grey-yellow to red-brown, 2–4 cm wide; the stem is red.

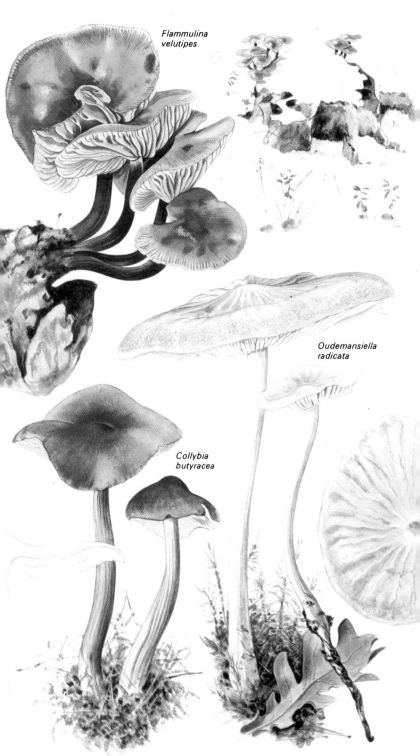

Flammulina velutipes

Oudemansiella radicata

Collybia butyracea

Marasmius oreades ⊙

(Bolt. ex Fr.) Fr. (from Greek *maraino*: dry, shrink and *oreias*: mountain nymph)

Fairy Ring Champignon

When fungi infest lawns and gardens it is hard to eradicate them. Some species, however, are both attractive and good to eat, such as Fairy Ring Champignon. It grows in rings, and the Latin name refers to ancient superstitions about 'fairy rings'. Commonly found at roadsides, and in parks, meadows and pastures, it belongs to a group characterized by tough and fibrous stems. Unlike other fungi, which soon rot or shrivel, it is tough enough to survive repeated drying. The umbonate cap is 2–5 cm across, brown-yellow to red-brown when damp, and brown-yellow to chamois-coloured when dry. The tough, straight stem is 3–7 cm tall. The fresh scent resembles cloves and the taste is nutty. This fungus makes an excellent seasoning, is also very good fried and is suitable for drying. It grows from June to November, and is very common throughout the country.

The genus *Marasmius* includes a large number of small delicate species, which are often seen in great quantity. A species with a characteristic odour of garlic is *Marasmius scorodonius* ⊙ (Fr.) Fr. (from Greek *scorodon*: garlic). It grows in groups on decaying matter in the sunny open parts of woods and meadows.

A carpet of fallen pine needles is often covered in *Micromphale perforans* ⊗ (Hoffm. ex Fr.) S. F. Gray (from Latin *perforans*: penetrating). Grey-brown, 2–3 cm tall, with a thin, radially-pleated cap, it is one centimetre across with a thin velvety stem. Each fungus grows on a separate pine needle.

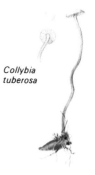

Collybia
tuberosa

Another species is the pretty white Little Wheel Toadstool *Marasmius rotula* ⊗ (Scop. ex Fr.) Fr. (from Latin *rota*: wheel). Its gills grow together at the stem to form a collar. It grows commonly on twigs and branches. A species that is usually found on fallen leaves is *Marasmius epiphyllus* ⊗ (Pers. ex Fr.) Fr. (from Greek *epi*: on and *phyllon*: leaf). The cap is 0.5–1 cm broad, membrane-like and pure white. The stem is white above, brown below.

In certain Tough Shanks the stem grows out of a knob-shaped formation or sclerotium. One of these species is *Collybia tuberosa* ⊗ (Bull. ex Fr.) Quél. (from Latin *tuber*: knob). This has purple-coloured sclerotia. Another one, this time with red-brown sclerotia, is *Xeromphalina cauticinalis* ⊗ (With. ex Fr.) Kühn & Maire (from Greek *kaulos*: stalk).

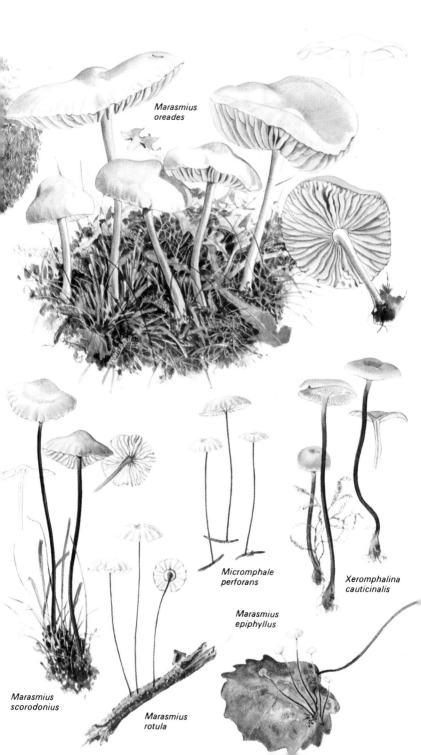

*Marasmius
oreades*

*Micromphale
perforans*

*Xeromphalina
cauticinalis*

*Marasmius
epiphyllus*

*Marasmius
scorodonius*

*Marasmius
rotula*

Amanita muscaria

(L. ex Fr.) Pers. ex S. F. Gray (from Greek *amanites*: a name used in the second century A.D. for an edible mushroom, later transferred to this genus by the Dutch scholar Persoon, and Latin *muscaria*: fly)

Fly Agaric

One of the most beautiful sights of autumn is a group of fiery red Fly Agarics shining in the sun, brilliant against the damp green grass and the fallen yellow leaves of the birch. The red Fly Agaric has become almost a symbol for all fungi: it decorates children's books, toys and Christmas cards. However, at one time in many parts of Europe and Asia, the Fly Agaric was associated with demons and evil, and with insects, especially flies, which were surrounded by an aura of superstition. It is not clear whether the connection with flies gave the fungus its name, or whether it was the discovery that one of its extracts can kill flies. Linnaeus knew that Fly Agarics could be used as an insecticide, and recounts in his *Journey through Skåne* how, by smearing the pulp of rotten Fly Agarics on walls: 'The beastly things die as if the plague had come among them.' Linnaeus also gave the fungus the name *Agaricus muscarius*. Some two hundred years later researchers isolated the substance, the insecticide ibotenic acid.

The Fly Agaric was once thought to be a deadly poison but it is now considered less poisonous. The fungus contains relatively little of the poison muscarin, which attacks the nervous system, but it does, however, have an hallucinogenic effect. Three active substances have been isolated: muscinol, muscazon and ibotenic acid. The hallucinogenic effect has been described in accounts of people living in Siberia, who use Fly Agarics as an intoxicant.

The main features of the genus *Amanita* are free gills and the fact that the developing fruitbodies are surrounded by two veils, an outer and an inner. The outer remains in the form of scales on the cap and a volva at the base of the stem. The inner veil forms a ring around the stem. The Fly Agaric is frequently found in deciduous and coniferous woods, and forms mycorrhizal associations with both birches and pines.

Amanita muscaria var. regalis (Fr.) Sacc. (from Latin *regalis*: royal) is very similar but yellow-brown to tobacco brown or grey-brown. It also may be found in deciduous and coniferous woods but is less common than the typical variety.

Amanita muscaria

Amanita muscaria
var. regalis

Amanita pantherina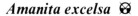

(DC. ex Fr.) Kummer (from Latin *pantherinus*: panther)

The Panther

Amanita rubescens

One of the fungi which is poisonous, but not normally deadly. It grows in deciduous and coniferous woods during summer and autumn, and in some years is fairly common. The cap is 8–12 cm across, olive grey to pale grey-brown, olive brown or sooty brown, often with a striate margin. The remnants of the veil are white and fall off easily. The stem, 5–10 cm high, is white, shiny and silky with a thin irregular ring easily torn off. The stem is wider at the base with a volva of two or three concentric rings. The scent is like raw potatoes. The rings and the remnants of the veil are easily washed away by rain, and the fungus can then be confused with the edible 'ringless' species. Serious poisoning cases have been caused by confusing this fungus with the 'ringless' species, or with species of *Agaricus* and *Macrolepiota*. The nature and effect of the poison are still not fully understood.

Amanita excelsa

(Fr.) Kummer (from Latin *excelsus*: lofty, tall)

Poisonous, suspect, edible: these are the judgements of different writers on this stout, beautifully-marked *Amanita*. It springs up early in the summer, often growing in groups, in both deciduous and coniferous woods. The light grey to grey-brown cap is hemispherical at first, and then expands. It is 10–15 cm broad and often has many remnants of the veil. The upper part of the stem is white, the lower part off-white to grey with a touch of brown. It has a swollen base with a volva reduced to scales. The white ring is distinctly striated. The scent is mildly earthy, and the taste mild, but the uncertainty about the edibility of this fungus means that it should definitely be avoided. This applies to *Amanita* in general. Even the Blusher *Amanita rubescens* (Pers. ex Fr.) S. F. Gray (from Latin *rubescens*: reddening) should be treated with caution, even though in some places it is considered good to eat. Anyone who wants to try it should always boil it first. The Blusher is red-grey to red-brown. The stem is up to 15 cm high with a striate ring and a scaly non-volvate base. The flesh reddens when touched. The scent suggests earth or raw potatoes. The Blusher, one of the more common species, grows in copses, coniferous and deciduous woods as early as July, but is also seen in the autumn.

Amanita pantherina

Amanita excelsa

Amanita virosa

(Lam. ex Fr.) Quél. (from Latin *virosus*: stinking, poisonous)

Destroying Angel

This species is deadly poisonous. There is a considerable risk of confusing the Destroying Angel with white edible mushrooms, especially in the button stage. In woodland areas be very careful in picking white fungi which are in an early stage of development. There have been cases where even people with a good knowledge of fungi have been poisoned.

It is characteristic of *Amanita* in having two veils. The outer veil is firmly attached to the stem, and forms a broad membranous volva when it breaks. There are few or no remnants of the veil on the cap in this species. The gills are revealed when the inner veil comes away from the cap margin and remains as a torn ring around the stem. An edible mushroom has a well-developed ring, but generally no volva or rings of scales around the base. Therefore, always examine the stem.

The wholly white cap of the Destroying Angel is first conical, then bell-shaped, and is up to 10 cm across. The pure white gills have a notched edge. The spore print is pure white. The stem, often curved in a bow-like shape, has wavy scales and is 10–15 cm high. The fungus is slimy in wet weather, shiny in dry weather, and has a heavy, stifling, soporific smell – in young specimens like bad perfume, in older ones the smell is rank or stinking. It is found during August and September in sour, poor soil mainly under beeches in deciduous woods.

Note that this *Amanita* does not grow on open grassland like several of the mushrooms. Normally the species is fairly rare, but in certain years it can spring up in quantity. Alongside the Death Cap, this is our most poisonous fungus and it has caused a large number of deaths in Europe during recent decades. The poisons in this fungus are almost the same as in the Death Cap (see p. 56), but their effects have not been as closely studied.

The gills of Amanita virosa *are white*

Mushroom gills are first pink or red-grey, then dark brown from the dark spores

Amanita citrina

(Schaeff.) S. F. Gray (from Latin *citrinus*: lemon, lemon-yellow)

False Death Cap

This species can be very similar to certain forms of Death Cap, but it is smaller and has a pale lemon-yellow cap with brownish remnants of the veil. The gills are crowded and white. The stem swells at the base into an onion shape. In damp weather this fungus, which grows under pines and beeches, is sticky, and in dry weather it is silky and shining. The scent suggests raw potatoes and the taste is mild. Recent investigations have contradicted the earlier belief that it was poisonous, but it is best left alone.

Base of the stem of Amanita citrina

One of the smaller species is *Amanita porphyria* ⊠ (Alb. & Schw. ex Fr.) Gillet (from Latin *porphyreus*: purple). The thin, grey-brown to sooty-brown cap with a purple tinge does not usually have remnants of a veil. The slender stem has a ring which is first grey, then black, and is surrounded at the base by a loose and fallen volva. This fungus is poisonous. It occurs in coniferous woods in early autumn.

Amanita porphyria

Amanita phalloides 🕱🕱

(Vaill. ex Fr.) Quél. (from Latin *phalloides*: like *Phallus* [stink-horn])

Death Cap

Why are certain *Amanita* species deadly poisonous, and others not? How have the poisonous substances been divided among the species in the course of their evolution? How are living cells and tissues affected by poisons? How does the fungal environment influence the combination of its poisons? Such problems are now being carefully studied. The Death Cap's poisons – phalloidin and amanitin – are made up of a number of substances, known respectively as phallotoxins and amanitatoxins. Its toxins are probably responsible for 90 per cent of all deaths from mushroom poisoning. How the poisons work is discussed in the introduction to vol. 1.

The cap, sticky in wet weather, is first egg-shaped, later hemispherical and finally expands to a breadth of 6–12 cm. The outer veil remains as a saccate volva around the base. The cap colour varies from pale yellow to olive green, grey-green or grey-brown and the gills from white to yellow-white. The stem is 10–15 cm tall, of even thickness, with a well-defined ring which is striped at first. The stem is white with a green to grey wavy pattern. The sweet and honey-like scent turns rank in older specimens. The Death Cap grows during August and September in rich soil where there are plenty of plants and herbs: deciduous meadows and woods and in parks, especially under oaks. A widespread species.

Amanita citrina

Amanita phalloides

Amanita vaginata ⊖

(Bull. ex Fr.) Vitt. (from Latin *vagina*: sheath, scabbard)

Grisette

The genus *Amanita* contains a group of species which do not appear to have a ring. Sometimes these are treated as one species, sometimes as several. They were formerly placed in a genus of their own, *Amanitopsis* (= like *Amanita*). Actually these fungi do form a ring, but it is not firm enough to hold together as the fruitbody develops.

The various types of *Amanita vaginata* all have a thin cap which is striate at the margin, a tall, fairly narrow stem and a well-developed volva of irregular height at the base. The different types (forms, varieties or species, according to various scholars) differ from each other mainly in the cap colour. It can be almost white, light grey, grey-brown, light brown or yellow-brown. All have a mild scent and taste and are edible. The grey form is often said to be good to eat. One should be cautious, however: it is easy to confuse these with other species which have lost their ring. Grisettes grow in soil where there is plenty of leaf mould, often in deciduous woods, mixed woods and in copses. They are widespread throughout Britain.

A larger and coarser type, often called the park form, is clearly distinct from the others. It is now treated as an individual species: *Amanita inaurata* ⊖ Secr. (from Latin *inauratus*: with gold). An imposing fungus, up to 25 cm tall, the cap is yellow-brown to grey-brown, clearly striate at the margin and up to 15 cm across. The stout stem has a wavy grey pattern. The volva is distinct in young specimens, but it soon vanishes almost completely, leaving rings of scales around the base. Unlike other ringless *Amanitas*, this species generally retains wart-like fragments of its veil on the cap. It is occasionally found in parks and copses during late summer and early autumn and can be confused with the *Amanita rubescens* (see p. 52) and *Macrolepiota rhacodes* (see p. 62), both of which can sometimes lose their rings.

Even though the ringless *Amanitas* described here are edible, and *A. inaurata* is even considered excellent eating, there is good reason for caution. It is worth considering what the mycologist Waldemar Bülow wrote about all of them at the beginning of this century: 'They belong to the considerable number of fungi to which it is better to devote a purely botanical interest than to give them a place on the table.'

*Amanita
vaginata*

*Amanita
inaurata*

Volvariella speciosa ⊗

(Fr.) Singer (from Latin *volva*: sheath and *speciosus*: beautiful, stately)

Rose-gilled Grisette

Amanitas, with their very special structure, have few close relatives, but it is thought that the pink-spored species of *Volvariella* and *Pluteus* are nearest to them. *Pluteus* lacks veils entirely whereas *Volvariella* has an outer veil, the remains of which form an irregular volva at the stem base. There are some ten species in Britain but most of them are rare.

Volvariella speciosa is a relatively large species, with a cap which is first bell-shaped, expanding to 6–10 cm across. It is white to brown-grey, and sticky when damp. The gills are white at first, but are then coloured light pink to salmon pink by the spores. In its early stages of development this fungus looks like an egg. The stem is 8–10 cm high. The flesh does not smell at first, but soon develops a musty disagreeable scent. The fungus is considered edible, but cannot be recommended. It can be confused with the ringless *Amanitas*, but these have a striate margin to their caps and their gills remain white. It usually grows among rotting straw, on compost heaps and on heavily-manured pastures in early summer.

Most *Volvariella* species are smaller. Some grow on the stumps and trunks of deciduous trees and one grows on rotting *Clitocybe* species. Those found in meadows and pastures may possibly be confused with *Agaricus* species. One such species is shown in the margin.

Pluteus cervinus ⊗

(Schaeff. ex Fr.) Kummer (from Latin *pluteus*: roof, shield and *cervinus*: fawn-coloured)

Fawn Pluteus

In Britain there are thirty-six *Pluteus* species. Most are small and grow on decaying wood and stumps. As with *Volvariella*, the cap can easily be separated from the stem, which is usually narrow, straight and rather rigid.

The Fawn Pluteus cap is grey to brown or blackish brown, almost flat and up to 12 cm wide. It is innately radially fibrillose. The crowded gills are broad at first, grey-white and later salmon pink. The stem is 10 cm tall, grey-white with darker fibrils. The scent suggests potatoes and the taste is mild. This fungus grows commonly on sawdust and the stumps of deciduous trees during late summer and autumn.

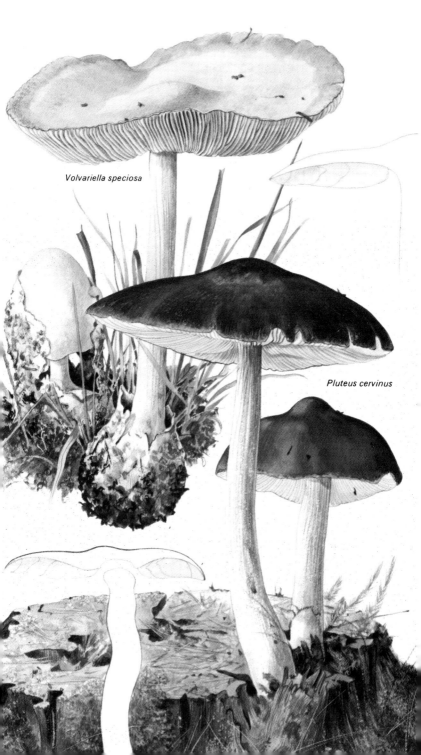

Volvariella speciosa

Pluteus cervinus

Macrolepiota procera ⊖

(Scop. ex Fr.) Singer (from Greek *makros*: large and *lepis*: scale, plus Latin *procerus*: tall, high)

Parasol Mushroom

To stand in a well-tended pasture, heavy with scent after warm summer rain, and to catch sight of a group of Parasol Mushrooms in a sunlit glade can be a rich experience. This stately fungus – it may be 30–40 cm tall and more than 30 cm wide – can hardly be confused with any other. It has long been highly regarded for the table and is one of the best edible fungi. It is quite common throughout Britain, occurring in deciduous woodland, outskirts of copses and occasionally in pastures. The Parasol Mushroom is a gregarious species, sometimes forming rings. Colonies of thirty, even of a hundred, have been recorded, but it is more usual to find it in groups of less than ten.

Macrolepiota excoriata

    When young the Parasol Mushroom looks like a grey-brown drumstick. As it grows, the upper ball-shaped part of the club develops into the bell-like to flat cap with grey-brown scales and a brownish umbonate centre. It is usually 20–30 cm wide. The stem is dark and scaly, and has a ring and a broad swollen base. The scent is pleasant and the taste nutty. Young specimens are best fried but they can also be eaten raw. The stem below the ring is tough, and this – together with older specimens – should be avoided. The growing season is from July to October.

Macrolepiota rhacodes ⊖

(Vitt.) Singer (from Greek *rhakodes*: broken)

Shaggy Parasol

This species, less common than the Parasol Mushroom, also belongs to the genus *Macrolepiota*, the scaly agarics with a movable ring. It grows in soil that has been cultivated, often on compost in parks and gardens, but sometimes in the carpet of needles under pine trees and on ant-hills, often forming large rings. The cap has more pronounced scales and less of an umbo than the Parasol Mushroom and is seldom more than 15 cm across. The stem is whitish and, unlike the Parasol Mushroom, lacks scales and has a greatly distended base. It grows up to 20 cm tall. The flesh reddens when broken, the scent is a trifle sour and the taste is rather sharp, but it makes excellent eating if properly prepared. It grows from August to October.

    Another species within this group is *Macrolepiota excoriata* ⊖ (Schaeff. ex Fr.) Singer (from Latin *excoriatus*: flayed) which grows on pastures in summer and autumn, but is found only occasionally.

Caps of Parasol (above and Shaggy Parasol Mushrooms

Macrolepiota procera

Section of Macrolepiota rhacodes

Macrolepiota rhacodes

Lepiota cristata

(Bolt. ex Fr.) Kummer (from Latin *cristatus*: crested)

Stinking Parasol

The *Lepiotas* were formerly placed in one large white-spored genus containing both large and numerous smaller species. One recognizable field characteristic of this group is the more or less pronounced firmly-attached scales or granulose fragments on the cap, and in certain species on the stem. Also, the stem has a ring but never a volva at the base. The species shown on the previous page belong to a natural group of some ten species, but most of the remaining species – about fifty in Europe – belong to the genus *Lepiota*. They have a fixed ring, which often disappears, and free gills. Several are very poisonous. A number of them, such as *Lepiota cristata*, have an unpleasant smell. Found on lawns or in leaf-mould in parks and gardens, often in rings.

The cap is at first bell-like but soon expands. It is white with a brown centre of red-brown granulose scales, 2–5 cm wide, and lacks an umbo. The gills are thin, crowded and white. The fragile white stem, often with a pinkish tinge, grows to 6 cm high. The ring is white, thin, and soon disappears. The scent recalls coal-gas or rubber. This fungus, found from August to October, is not edible. Closely related species with a reddish or lilac tinge and dark scales have caused serious cases of poisoning, some of them fatal.

Another species, uncommon in Britain, is *Lepiota clypeolaria* ✪ (Bull. ex Fr.) Kummer (from Latin *clypeus*: shield). It grows in coniferous and deciduous woods. The cap varies from bell-like to broad and flat. It is whitish yellow with brown scales, 4–8 cm across. The light brown stem has widely-spaced scales, becoming uneven and felty or downy towards the base. It is 4–10 cm high. The thin ring soon disappears. The scent is slight, the taste mild. This fungus, found in August and September, is edible but should be avoided as it could be confused with poisonous species.

Cystoderma amianthinum ✪

([Scop.] Fr.) Fayod (from Greek *kystis*: blister and *derma*: skin, plus Latin *amianthinus*: amianth [asbestos])

Saffron Parasol

The genus *Cystoderma* is quite distinct from *Lepiota*. The gills are widely spaced, the granulose outer veil consists of round cells and the ring points upwards. Several species are small, up to 5 cm wide at most. One of the commonest is *Cystoderma amianthinum* which grows from August to October. Another common species is *Cystoderma carcharias* ✪ (Pers. ex Secr.) Fayod (from Latin *carcharias*: shark). Both have an unpleasant smell and are inedible.

Cystoderma species grow among grass and moss in deciduous and coniferous woods

Lepiota cristata

Lepiota clypeolaria

*Cystoderma
amianthinum*

*Cystoderma
granulosum*

*Cystoderma
carcharias*

Agaricus campestris ⊖

L. ex Fr. (from Greek *agarikon*: name of a fungus which was probably eaten by a hunting community from Agaria on the lake of Azovska, plus Latin *campestris*: field)

Field Mushroom

In France *champignon* means fungi in general. Cultivated mushrooms, among the most highly-prized fungi, are *champignons de couche* (bed fungi). But the danger of confusing them with deadly poisonous *Amanitas* must be constantly stressed. The only way to avoid this confusion is to study carefully the features of the genus *Agaricus*. The cap can be separated from the stem without damage. The gills are free from the stem. They are whitish at first, then become pink or pinkish-grey and finally dark chocolate brown or almost black with purple-brown spores. The stem has a ring which may be single or complex, but no volva at the base. Although the genus is well defined, the species can be very hard to separate. In Europe there are probably about fifty, divided into ten groups.

Agaricus
bisporus (wild form)

The Field Mushroom, which belongs to a group of species with reddening flesh (section *Rubescentes*), often grows in meadows and pastures, and can appear in masses during the warm August rains after a dry, warm summer. The cap is first hemispherical, then slightly arched, finally almost flat, 5–10 cm wide. The cap skin is white, varying from silky or downy to delicately scaly. The gills are pink to meat red, then red-grey and finally very dark brown. The white stem is 3–7 cm high and has a single, thin, white, often barely perceptible ring. The scent is slightly sour but pleasant, and the taste mild. The Field Mushroom has a delicious taste and grows from July to October throughout the British Isles.

Below: spore print

The cultivated mushroom *Agaricus bisporus* ⊖ (J. Lange) Pilát (from Latin *bisporus*: two-spored), and a related brownish species, upon which the mushroom industry was based, were formerly thought to be forms of the Field Mushrooms. However, they belong to quite another group, distinguished among other factors by a complex ring. They are easy to cultivate because they thrive in manure. In the same group there is *Agaricus bitorquis* ⊖ (Quél.) Sacc. (from Latin *bitorquis*: adorned with two necklaces) a common species along roadsides and in parks.

Left: single ring
Right: double ring

A fourth species with reddening flesh is the large and solid *Agaricus bernardii* ⊛ (Quél.) Sacc. (from the name of the French scholar G. Bernard). It grows up to 15 cm wide. In dry weather the cap skin breaks into widely-spaced scales. This fungus belongs to the same group as the cultivated mushroom, but it has an unpleasant smell and is not edible. It grows in meadows near the sea and along roadsides from June to September.

Agaricus
bitorquis

Agaricus campestris

Agaricus bernardii

Agaricus bisporus

Agaricus arvensis ⊖

Schaeff. ex Fr. (and related species) (from Latin *arvensis*: from fields)

Horse Mushroom

This is a group name for several species which turn yellow when rubbed – a distinctive feature of the second main section of mushrooms *Flavescentes* (from Latin *flavescens*: yellowing). Horse Mushrooms have an almond or aniseed smell and a complex ring called the 'cog-wheel'. Often there are scales on the underside of the ring.

Elias Fries' Swedish name for the Horse Mushroom was the Common Champignon. There have been many interpretations of Fries' *arvensis*. The species is not found in woodland, and Fries writes: 'it is found most in the open field, also in cultivated soil'. The species fruits as early as June if the weather is damp and warm. Other related species are autumn fungi which grow in pine litter or among leaves in beech woods.

There are some features which are common to all Horse Mushrooms. The white, dry, silky cap is never slimy. It is at first irregularly round like a snowball and then becomes convex or flattened, 10–15 cm wide. The gills, revealed when the ring separates from the margin of the cap, are red-grey at first, but become very dark brown. The white stem is somewhat distended at the base. This fungus has an outstanding flavour and can be eaten raw.

Base of
A. abruptibulbus (top)
and A. silvicola

Agaricus abruptibulbus ⊖ Peck (from Latin *abruptus*: crosswise, abrupt and *bulbus*: onion), often found in coniferous woods, deeply rooted in pine litter, differs in having an abruptly swollen, marginate base. A third species *Agaricus silvicola* ⊖ (Vitt. ex Fr.) Sacc. (from Latin *silvicola*: growing in woods) has a thinner ring with a barely perceptible 'cog-wheel' and a club-shaped swelling at the base.

In another group of 'yellowing' species there is the large *Agaricus macrosporus* ⊖ Møller & Schaeffer (from Greek *makros*: long and *spora*: seed) which grows up to 25 cm wide. It also differs from the Horse Mushroom in that the underside of the ring is floccose and the smell of fully-grown specimens is unpleasant.

Agaricus xanthodermus ⊗

Genevier (from Greek *xanthos*: yellow and *derma*: skin)

Yellow-staining Mushroom

This fungus resembles the Horse Mushroom but usually has a disagreeable carbolic-like smell. The species is poisonous to some and should be avoided. It stains bright yellow especially at the stem base, when bruised. It grows in parks and gardens, often in clusters loosely rooted in leaf-mould and compost, and is fairly common in Britain.

Agaricus
xanthodermus

68–69

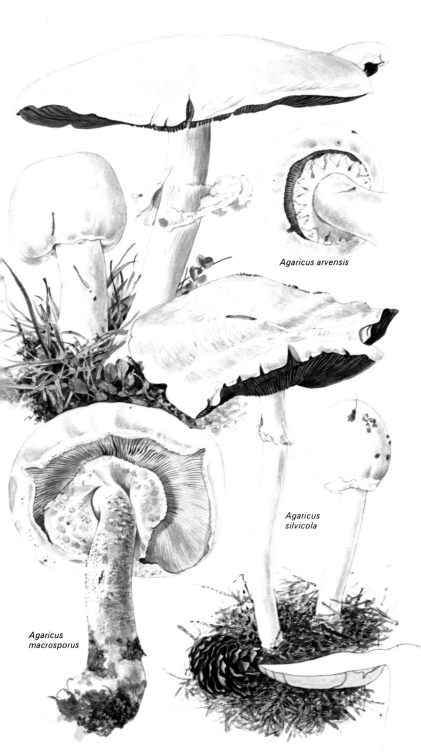

Agaricus arvensis

Agaricus silvicola

Agaricus macrosporus

Agaricus augustus ♀

Fr. (from Latin *augustus*: inviting respect)

This stately mushroom certainly justifies its Latin name. The straw-coloured cap with brown scales grows to 25 cm wide, and the white stem is often up to 20 cm high. It grows under deciduous trees, especially oak, and in the carpet of needles under old pines, in open, warm pastures and woodland clearings, often on chalky ground. It is not common.

This species belongs to the 'yellowing' group, section *Flavescentes* (see p. 68), and is one of the few to have brownish scales. The underside of the ring is floccose and fissured, unlike other brownish species. The cap is at first semiglobate, becomes bell-shaped with a flat centre and then is completely flat. The gills are white for a long period and change to chocolate brown, never pink or red. The stem is often floccose below the ring, which in older specimens hangs downwards. The pleasant scent suggests almond or aniseed, and the taste is exquisite. *Agaricus augustus* grows from July to October.

A unique group of yellowing mushrooms is the section *Minores*, the dwarf mushrooms. Many are 5 cm wide at most and thus are very different from *Agaricus augustus* and other large species, the *Majores*. A species of dwarf mushroom is illustrated on this page.

Dwarf mushroom

The underside of the ring on Agaricus augustus *is floccose and fissured*

Agaricus silvaticus ♀

Schaeff. ex Fr. (from Latin *silvaticus*: wood)

Wood Mushroom

The majority of brown-scaled mushrooms belong to the 'reddening' group *Rubescentes*. The *silvatica* group are woodland fungi. The species within this group are especially common in pine woods on chalky ground, and often grow in clusters, deeply rooted in the pine litter. Many forms have been described, but they are hard to distinguish.

The Wood Mushroom is fairly small, 5–10 cm wide, and reddens slightly when broken. The cap, white-grey with small nut-brown scales, is first hemispherical and then slightly convex. The gills are first pale grey-brown, then chocolate brown. The stem is white-grey, up to 10 cm tall, and swollen at the base. The greyish ring is fairly stiff with a white-scaled underside, and quickly begins to hang downwards. The scent is slightly sourish, but the taste is mild and good. Grows from August to October.

A related specimen is the large sturdy *Agaricus langei* ♀ Møller (from the name of the Danish scholar J. Lange) which has pronounced scales and reddens markedly.

Agaricus augustus

Agaricus silvaticus

Agaricus langei

Coprinus comatus ☉

(Müll. ex Fr.) S. F. Gray (from Greek *kopros*: droppings, and Latin *comatus*: hairy)

Lawyer's Wig, Shaggy Cap

Ink Caps acquired their name because the gills of older specimens dissolve into a black fluid. This 'ink' does not, however, help to spread the spores which are disseminated by the wind before the gills autodigest. Ink Caps grow very quickly, and in summer and autumn they fruit in great quantity, especially after rain. Normally they grow in parks and gardens, on refuse dumps and along roads.

Lawyer's Wig is large and pretty. Young undeveloped specimens look like eggs half-hidden in the grass. Later the fungus becomes rather cylindrical, like a narrow egg, and then expands into a bell-like shape. The cap is white to whitish brown, up to 15 cm high, with pronounced scales. The top of the cap is light brown. The gills are at first white, then rose pink and finally black and – like the cap – form an inky fluid. The white stem has a narrow ring which falls away. The scent and flavour are pleasant. Young specimens can compete with mushrooms as delicacies for the table. However, avoid specimens that have grown on land that has recently been heavily manured. The Lawyer's Wig is common throughout most of Britain in places where there is plenty of humus.

Another Ink Cap of similar size and shape *Coprinus picaceus* ☻ (Bull.) Fr. (from Latin *picaceus*: pitch) grows in beech woods. It is first enclosed in a veil which splits and leaves large scales on the cap. The cap is brown-grey to browny black or pitch black. This species is not edible.

Coprinus picaceus

Coprinus atramentarius ☉ (⚥ *with alcohol*)

(Bull. ex Fr.) Fr. (from Latin *atramentum*: ink)

Common Ink Cap

A very common species frequently found in clusters near old stumps and tree roots on lawns. The fungus is bell-shaped, grey to greyish black with touches of brown. The light grey gills blacken with age. The stem is short and grey. Young specimens are good to eat, but this fungus produces ill effects including nausea and vomiting if taken with alcohol. These effects are the same as those of antabuse and other medicines used in the treatment of alcoholism.

Around stumps and trees one often finds clusters of another rather smaller, light yellow-brown to brown ink fungus: the Glistening Ink Cap *Coprinus micaceus* ☻ (Bull. ex Fr.) (from Latin *micaceus*: grainy). The furrowed cap has small glistening grains over its entire surface which soon disappear. This species is inedible.

Coprinus micaceus

Coprinus comatus

Coprinus atramentarius

Coprinus disseminatus ⊗

(Pers. ex Fr.) S. F. Gray (from Latin *disseminatus*: dispersed)

Trooping Crumble Cap, Fairies' Bonnets

One of the more common and easily recognized small Ink Caps is the Trooping Crumble Cap, once placed in the genus *Psathyrella* (from Greek *psathyros*: fragile). As early as spring you can see it growing in tight clusters on the decaying stumps of deciduous trees. The cap is bell-shaped and deeply furrowed at the edge, 0.5–1 cm wide. The colour is clayey grey to yellow-brown or dark brown. The gills are lilac brown to grey-brown, and are not deliquescent. The short thin stem has a fine down. Normally grows on rotten wood, but also may be found in rich soil.

There is a rich and interesting flora of fungi to be found on animal droppings, including several agarics as well as a host of ascomycetes and other microscopic species. Many fungi are only found in the droppings of certain animals. One of the agarics most often seen is the Grey Mottle Gill *Panaeolus sphinctrinus* ⊗ (Fr.) Quél. (from Greek *panaiolos*: irregularly coloured, and *sphincter*: band). It has a bell-like cap 1–2 cm broad, with a somewhat fringed edge. Its colours are varying shades of brown. The gills are mottled black and brown – mottled because the spores mature at different rates, hence the name of the genus. The stem is tall and straight. This fungus usually grows on cow-pats.

Psilocybe coprophila

Psilocybe coprophila ⊗ (Bull. ex Fr.) Quél. (from Greek *psilos*: naked and *kybe*: head, plus Latin *coprophilus*: dung-loving) is also found on dung or heavily manured soil. It has a shiny hemispherical cap, 2–3 cm broad, which is light to dark brown in colour. The sticky skin of the cap can be easily peeled off. The stem is white-brown above, light brown below. The species is fairly common. The genus *Psilocybe* has recently been the subject of a great deal of research by chemists, doctors and botanists. It is now clear that certain species contain substances that produce an hallucinogenic effect, one of which is psilocybin (see vol. 1, p. 31).

Another coprophilous species is Yellow Cow-pat Toadstool *Bolbitius vitellinus* ⊗ (Pers.) Fr. (from Greek *bolbiton*: cow-dung, and Latin *vitellus*: egg-yellow). The sticky cap has a furrowed margin 2–6 cm across when fully expanded. The colour is at first egg yellow, becoming straw yellow or dirty yellow. The gills are pale brown. The stem is whitish yellow, with a granular surface. This fungus is fairly common throughout northern Europe during summer and autumn, often growing on cow-pats, but also on rotting wood, around haystacks and on compost heaps.

Coprinus disseminatus

Panaeolus
sphinctrinus

Bolbitius
vitellinus

Stropharia hornemannii

(Weinm. ex Fr.) Lundell & Nannfeldt (from Greek *strophos*: belt, girdle, and Hornemann, Danish botanist)

The genus *Stropharia* contains species with a collar-like ring, a dark violet spore print and gills widely spaced at the stem. They are slimy when damp. Most are fairly small, normally growing in soil which is rich in mould, on dung, or on rotten wood. *Stropharia hornemannii* is the largest species – a stout handsome fungus which grows in or beside mossy swamps or peat bogs in coniferous woods. The cap is 6–12 cm wide, brown-grey to greyish red with a touch of violet. The gills are first grey-red, then brownish violet. The stem is white to dirty yellow, 8–12 cm tall, and granulose below the collar-like ring, which is often torn. The ring becomes violet from the spore deposit. This fungus smells and tastes disagreeable and is not edible. It occurs only rarely in Britain, but is more common elsewhere in Europe.

Stropharia aeruginosa

(Curt. ex Fr.) Quél. (from Latin *aeruginosus*: verdigris-coloured)

Verdigris Toadstool

Its verdigris-like colouring is striking and always arouses the attention and interest of the woodland walker. The cap is 5–10 cm in width, convex and somewhat umbonate, with a ragged white margin and a layer of slime which later disappears. Its colour pales with age to greeny yellow or yellowy beige. The stem is 5–8 cm tall, floccose and whitish green, with an upright ring at first, coloured a dark violet by the spores. The scent suggests radish, the taste is strong and unpleasant and the fungus is inedible. The Verdigris Toadstool occurs commonly throughout Britain between July and November. Various other greenish *Stropharia* species have been observed in recent years.

Stropharia semiglobata

Stropharia semiglobata

(Batsch. ex Fr.) Quél. (from Latin *semiglobatus*: hemispherical)

Dung Roundhead

This tall slender fungus is frequently found in meadows and on dung in pastures. The regular hemispherical cap is 1–3 cm broad and dirty yellow in colour. The narrow grey-yellow stem grows 8–15 cm high. The ephemeral ring is coloured by the falling spores and appears as a violet, ring-like film on the upper part of the stem.

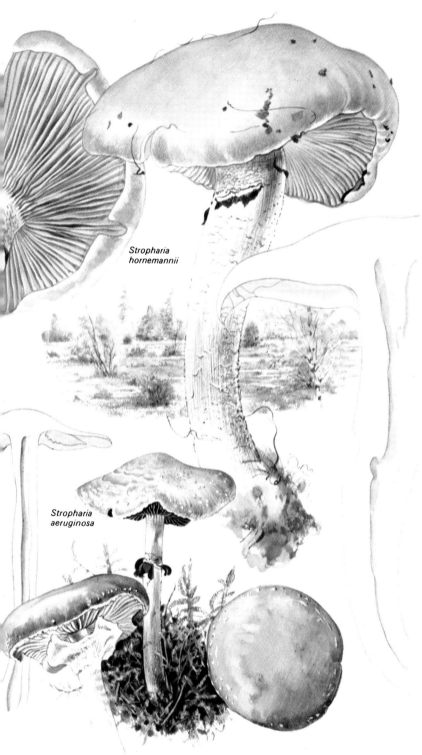

*Stropharia
hornemannii*

*Stropharia
aeruginosa*

Hypholoma capnoides ⊖

(Fr. ex Fr.) Kummer (from Greek *hyphe*: net and *loma*: fringe, plus *kapnos*: smoke)

The bright yellow *Hypholoma* species can be seen from afar. During a normally damp autumn they occur in crowded clusters on the stumps of trees. Various shades of yellow mingled with other colours are dominant. Another feature of the genus is the reduced fibrous or membranous veil. Part of the veil remains on the margin of the cap and part remains as fibrils on the upper part of the stem. The spore print is dark brown with a tinge of violet. One of the species is *Hypholoma capnoides* whose pale yellow to browny cap is 2–6 cm across and somewhat slimy when damp. The gills are smoky grey with a touch of violet and the stem is 5–10 cm high, shining, silky and pale yellow. The scent and taste are mild, and the fungus is edible. It usually grows on conifer stumps, especially pines.

Sulphur Tuft *Hypholoma fasciculare* ☒ (Huds. ex Fr.) Kummer (from Latin *fascis*: bunch, cluster) is found on deciduous trees. The convex cap is 2–6 cm across, sulphur yellow with a light brown centre and a whitish yellow margin, with fragments of the veil. The gills are at first sulphur yellow to yellow-green, later violet-brown. The curved stem is rust-coloured towards the base. The sulphur yellow flesh tastes bitter and smells unpleasant. The fungus is poisonous. This species is very common throughout northern Europe, and grows from May to November on stumps, roots and trunks, often in very large clusters.

*Hypholoma
fasciculare*

The brick red *Hypholoma, H. sublateritium* ⊗ (Fr.) Quél. (from Latin *sub*: in this context 'fairly', and *lateritius*: brick red) is a somewhat more robust species, with a cap up to 8 cm across. The cap shades from light brick red to red-brown, and the gills are yellow to yellowish grey-green. The stem is 5–12 cm tall with the same shading as the cap, and has a conspicuous fibrillose ring zone. The flesh is yellow above, red-brown below. The taste is rather sharp and the fungus inedible. It is fairly common and grows throughout the autumn on the stumps and roots of deciduous trees.

Another dark-spored genus which sometimes has a veil is *Psathyrella*. The Fringed Crumble Cap *Psathyrella candolleana* ⊖ (Fr.) Maire (from Greek *psathyros*: fragile, and A. de Candolle, Swiss botanist) is thin and fragile, almost entirely white or white-grey, with grey to pinkish grey gills which are mature to dark brown. The cap margin often has jagged fragments of veil. Grows profusely in damp places, on and around stumps, on soil and in grass from early spring to autumn. It is one of the few fungi to be seen in quantity in early summer.

*Psathyrella
candolleana*

*Hypholoma
capnoides*

*Hypholoma
sublateritium*

Pholiota squarrosa

(Müll. ex Fr.) Kummer (from Greek *pholis*: scale, and Latin *squarrosus*: rough)

Shaggy Pholiota

Pholiota is the generic name for a number of brown-spored fungi usually found in clusters on stumps and trunks. Most of them have a distinct ring or scales on the stem and the cap is often scaly. Recent research has revealed that the genus *Pholiota* is not homogeneous, and several genera are now recognized. Two typical ones are described here. The Shaggy Pholiota grows in thick, often large, clusters at the base of deciduous trees or on their stumps. It is fairly common throughout the entire northern temperate zones, except in the very far north. The cap is 5–10 cm across, brassy yellow to yellow-brown with touches of rusty red and is covered with coarse, dark, concentrically-arranged scales. The stem is also scaly below the ring. The gills are straw yellow to yellow-brown. The flesh is yellow, firm, has a radish smell and a rather sharp taste, but the caps of young specimens are edible.

A related species *Pholiota flammans* ⊗ (Fr.) Kummer (from Latin *flammans*: burning) is smaller and brighter in colour – golden yellow to fiery yellow, with sulphur yellow tufts and scales. It grows on conifer stumps in autumn, but in Britain is common only in Scotland. It is not edible.

Kuehneromyces mutabilis ⊖

(Schaeff. ex Fr.) Singer & Smith (from the name of the French scholar Kühner, and Latin *mutabilis*: changeable)

Two-toned Pholiota

Throughout summer and autumn large dense clusters of more than a hundred of these fungi can grow on stumps, at the base of dead trees, on many kinds of deciduous trees and also on pines. The cap is 2–6 cm broad, smooth in damp weather and rust yellow to cinnamon brown. It is hygrophanous, i.e. it absorbs water easily. When the cap dries it becomes honey yellow to ochre yellow in the centre but the edge remains brown for a long time. The stiff and fibrous stem has a barely perceptible ring. Under the rust brown gills the stem is smooth and pale but below the ring it is scaly and dark. The scent and taste are mild and pleasant. The species is common throughout Britain. Two-toned Pholiotas are good to eat, but only the caps should be used. It is commercially cultivated in Japan, Germany, Hungary and some other countries. Blocks of wood injected with this fungus are sold for cultivation in home or garden.

Pholiota flammans

80–81

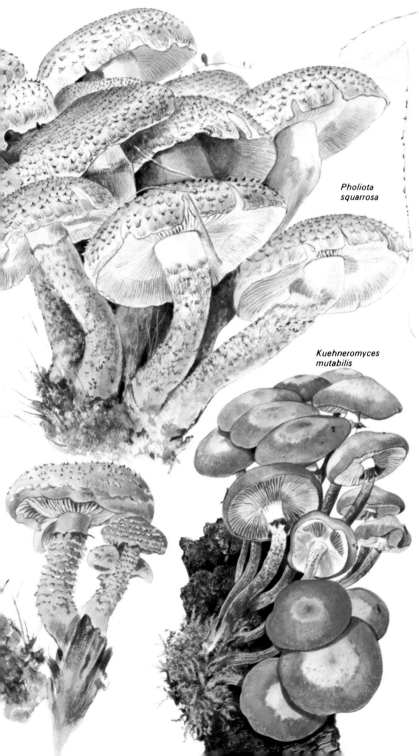

*Pholiota
squarrosa*

*Kuehneromyces
mutabilis*

Pholiota lenta ⊗

(Pers. ex Fr.) Singer (from Latin *lentus*: sticky)

Flammula (from Latin *flammula*: little flame), a group of brown-spored agarics, was formerly separated from *Pholiota*. It included species with a smooth cap, an ephemeral fibrillose veil and often a bitter taste. This genus now represents several smaller genera. Some of the species are placed in the genus *Pholiota* (see p. 80), which now includes agarics both with and without a distinct ring. Many species grow on wood, others on the ground. Among those that grow on the ground there is a group of species with a markedly slimy cap. One of these is *Pholiota lenta*, a fleshy, medium-sized fungus, fairly common during September and October in both coniferous and deciduous woods. The cap, 5–8 cm broad, is a pale clay shade with a tinge of green, often with flame-coloured flecks of veil. The gills are pale yellow to dark clay-coloured, thin and somewhat decurrent. The pale yellow stem, brownish towards the base, grows to 10 cm high. The scent is reminiscent of freshly-cut wood. The taste is radish-like but this fungus is not edible. In Britain and central Europe it grows mainly under beech trees.

Among the species which grow on the ground there is a group which grows amid the charcoal from fires. One is Charcoal Pholiota *Pholiota highlandensis* ⊗ (Peck.) Smith (from place-name). The cap is 2–3 cm across, yellow-brown to red-brown, smooth and slightly slimy, with floccose remnants of the veil. The pale yellow to olive brown gills are crowded and somewhat sinuate. The pale yellow stem has brownish scales, and grows 3–5 cm high. The scent is slightly radish-like and the taste mild. This inedible species grows during summer and autumn.

Pholiota alnicola ⊗

(Fr.) Singer (from Latin *alnicola*: on alder)

This yellow to rust-coloured species has a cap 3–8 cm wide with straw yellow to rust-coloured gills. The yellow to brownish stem does not have a ring. The scent is slight and the taste mild to slightly bitter. The fungus is inedible, as are all the other species in this group that grow on wood. It is occasionally found growing in small groups on alder.

Many brown-spored agarics have a small cap, up to one centimetre across. One such genus with species resembling *Mycena* (see p. 44) is the genus *Galerina* (from Latin *galerus*: fur cap). A *Galerina* species is illustrated on this page.

Pholiota lenta

Pholiota highlandensis

Pholiota alnicola

Inocybe fastigiata 🕱

(Schaeff. ex Fr.) Quél. (from Greek *inos*: fibre, thread and *kybe*: head, plus Latin *fastigiatus*: slanting)

The name for the *Inocybe* species comes from the appearance of the cap, in which the innate fibrils are radially arranged. The gills are grey at first, but later discoloured by the brown spores. Most species have a strong and special scent. Many of them smell strangely musty, suggesting earth, cellars or damp clothes. Nearly all are poisonous, and some extremely so, for example *I. patouillardii* 🕱🕱 Bres. (from the name of the French botanist Patouillard). This genus includes almost a hundred species in Britain.

Inocybe fastigiata has a yellow-brown to grey-brown rimose cap 3–8 cm broad, with a pointed umbo and an uneven margin. The gills are a dirty yellow-brown. The stem is hollow and 5–8 cm tall. The scent resembles that of fresh nuts. The fungus is poisonous. It grows fairly commonly during the autumn on pastures and in open woodland areas.

Another of the more common species is the Common White Inocybe *I. geophylla* 🕱 (from Greek *ge*: earth and *phyllon*: leaf). This is a small fungus with a silken sheen. It is usually off-white, but the colour can vary to a faint red and to violet. It smells strongly of damp earth. One of the more easily-recognized species is the chalk-loving *I. bongardii* 🕱 (Weinm.) Quél. (from the name of the Russian botanist Bongard). The veil makes the cap white at first, but later it is brown and scaly. The stem gradually becomes wine red to red-brown. When the flesh is cut it becomes carrot-coloured and smells strongly of pears. Another scaly species is the brown *I. lanuginosa* 🕱 (Bull. ex Fr.) Quél. (from Latin *lanugo*: fluff).

Inocybe bongardii

Hebeloma crustuliniforme 🕱

(Bull. ex Fr.) Quél. (from Greek *hebe*: young and *loma*: fringe, plus Latin *crustulum*: little cake, biscuit)

Fairy-cake Mushroom

These acrid agarics are another group of brown-spored fungi with a large number of species. They look rather like small *Tricholomas*. As the Latin name suggests, the veil is usually only seen in young specimens. The cap is 4–7 cm wide, sticky in wet weather, clay-coloured to light brown. The grey-brown gills weep dark droplets when damp, which leave dark marks at the edge when dry. The upper part of the dirty white stem has granulose scales. This species varies somewhat in shape and colour, has a strong radish-like smell, and is not edible. In the autumn it is common on heathland and in deciduous woods.

Inocybe lanuginosa

84–85

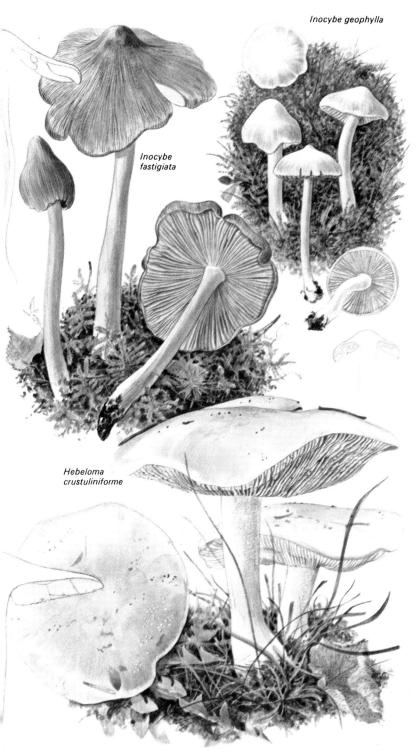

Inocybe geophylla

Inocybe fastigiata

Hebeloma crustuliniforme

Rozites caperata ⊖

(Pers. ex Fr.) Karsten (from the name of the French scholar Roze, and Latin *caperatus*: wrinkled, creased)

In Finland the fungus is called Granny's Nightcap. This fleshy, straw yellow to clay yellow fungus is easy to recognize because it looks as if it has been powdered with hoarfrost. It is a highly prized delicacy. The stem has a distinct ring, which in young specimens is thick, and connects the stem to the edge of the rounded cap. It seems to prefer soil that is poor in chalk, and is a mycorrhizal fungus, but not restricted to any particular kind of tree. It is common in the sparse beech woods of central Europe and southern Scandinavia, and in the coniferous woods of Scandinavia and Finland. In Britain it is extremely rare outside Scotland.

Rozites caperata

This fungus was formerly placed in *Pholiota* (see p. 80), although it is probably more closely related to species of the genus *Cortinarius* (see p. 88). Among the several features which justify a separate genus is its well-developed ring.

Rozites caperata is spread over large areas of the northern hemisphere, but is not found in the tropics or in the southern hemisphere. There are a number of species in Australia and New Zealand which are also placed in the genus *Rozites*, but they belong to another group of species, which are probably even closer to *Cortinarius* than our own representative of the genus.

The cap of *Rozites caperata* is at first rounded and grey-violet. Later it is convex to flat, straw yellow, wrinkled at the edge, up to 12 cm across. The light brown gills are adnate with a tooth. The white to yellow-grey stem is often somewhat streaky, and grows to 10–15 cm high. In older specimens the ring becomes tattered. The scent is slight and the taste mild, and it mixes very well with some of the spicy fungi, Chanterelles for example, but is also good with boletes, Russulas and milk caps. It is sometimes confused with certain *Agaricus* species but the risk of confusion with any poisonous fungus is not great. The species can be found from August to October.

Cortinarius triumphans ⊖

Fr. (from Latin *cortina*: curtain and *triumphans*: triumphal)

'To ignore *Cortinarius* is to dismiss 30 per cent of woodland fungi', wrote the author of a book on this genus. *Cortinarius* represents a large and variable group with at least four hundred species in Europe, but many are easily recognizable. They often resemble *Tricholomas* in shape, and some are among the prettiest fungi in our forests. *Cortinarius* gets its name from the fibrils, very similar to a spider's web, which form a cortinoid veil from the margin of the cap to the stem, concealing the gills of the young fungi. Other features of the group are the cinnamon-coloured spores, gills that change colour with age, and the joining of cap and the stem in such a way that they cannot be separated without damage. They are important because the majority form mycorrhizal associations with trees. In the next nine pages we describe some of the many species in this diverse and beautiful group, which has not yet been studied in depth.

Certain Phlegmacium *species have a bulbous marginate base*

Elias Fries, in his classic division of *Cortinarius*, separates a group of species which have a slimy cap and a dry stem in wet weather: the subgenus *Phlegmacium* (from Greek *phlegma*: slime). One of these is *Cortinarius triumphans*, a light yellow-brown species with ochre-coloured scaly garlands on the stem. The fleshy cap grows to 12 cm across. The gills are at first clay-coloured and hidden by a light yellow veil which soon splits apart and remains as fibrils on the stem and cap margin. Finally the gills and the veil take on the colour of the ochre-brown spores. The stem is of even thickness, somewhat swollen at the base and up to 15 cm tall. The scent is slight and the taste mild. The fungus is edible, and grows under birch trees from August to October.

Cortinarius glaucopus

Other members of the subgenus *Phlegmacium* include *Cortinarius glaucopus* ⊖ Fr. (from Latin *glaucopus*: blue-green stem). This is an olive yellow to grey-yellow radially striate species which grows in coniferous woods, both in chalky and in poor soil. Other species favour chalk. One of these is *Cortinarius varius* ⊖ Fr. (from Latin *varius*: changeable). This has a light yellow-brown cap, a white to whitish yellow stem and the gills are at first blue-violet. It grows in coniferous woodland.

A non-calcareous species of coniferous woods is *Cortinarius purpurascens* ⊖ Fr. (from Latin *purpurascens*: blushing purple) with a violet-brown cap and blue-violet gills which bruise reddish-violet. The flesh is bluish and the foot somewhat club-like. Other *Phlegmacium* species have a bulbous marginate base.

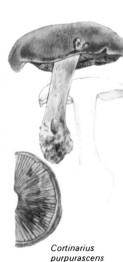

Cortinarius purpurascens

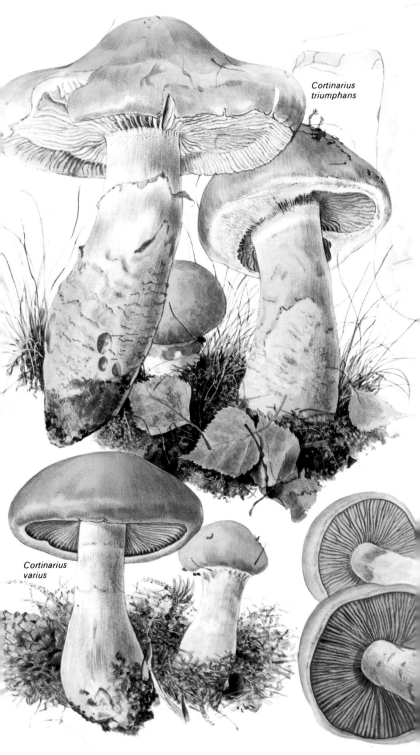

*Cortinarius
triumphans*

*Cortinarius
varius*

Cortinarius traganus

Fr. (from Latin *traganus*: smelling like a he-goat)

An important characteristic, useful in the identification of *Cortinarius* species, is the colour of the gills before the spores ripen and discolour them. Within the group *Phlegmacium* there are subgroups of species with whitish, yellowish, bluish or greenish gills. One example of a species with gills that are initially bluish is *Cortinarius varius*, which also has whitish and yellow-brown colourings (see p. 88). With other species the whole fungus, apart from the gills, may be more or less bluish. One of these is *Cortinarius traganus*, with a dry cap, which is now placed in the subgenus *Sericeocybe*. *Cortinarius traganus* occurs singly or in groups, growing mostly in coniferous woods on poor soil and is not uncommon in Scottish pine woods.

Two other bluish *Cortinarius* species are *Cortinarius camphoratus* ⊖ and *C. cyanites* ⊖. The northern limits of their distribution are not yet fully known. *Cortinarius traganus* differs particularly from other bluish species in the yellow-brown flesh and foul smell which suggests carbide gas. The silky shiny cap is grey-lilac and hemispherical at first with an inrolled margin. It then changes to browny yellow and becomes evenly convex, growing up to 10 cm across. The veil is grey-violet and the gills rust-coloured in young fungi. The grey-lilac stem, up to 12 cm high, widens into a club-like shape at the base. In older specimens the remnants of the veil and the gills are both coloured brown by the spores. The taste is unpleasant and the fungus is mildly poisonous. It grows from July to October. *Cortinarius camphoratus* Fr. (from Latin *camphoratus*: smelling of camphor) is very similar, but has lilac-coloured gills when young, and violet flesh. The smell is disagreeable, but does not suggest camphor – the name is misleading – as much as burned meat. It is inedible, and grows in moss in both coniferous and deciduous woods. There is, however, no authentic record for *C. camphoratus sensu* Fries in Britain.

Cortinarius cyanites Fr. (from Greek *kyanos*: dark blue) like one or two related species has bluish flesh which reddens. When species so closely resemble one another, it is only possible to make precise distinctions after careful study of both young and old specimens, and their manner of growth.

Cortinarius
camphoratus

Cortinarius traganus

Cortinarius cyanites

Cortinarius trivialis

Lange (from Latin *trivialis*: common, ordinary)

The outer veil which encloses the fungus embryo in *Cortinarius* can vary greatly. In one group of species – subgenus *Myxacium* (from Greek *myxa*: slime) – it is slimy on both cap and stem. This group includes at least twenty species in Europe, several of which are common and easy to recognize. In some species the coat of mucilage soon dries around the stem, and can then break up and form a pattern of 'garlands'. One such species is the brown-yellow to light olive-coloured *Cortinarius trivialis*, whose pattern resembles a spiral staircase or the markings of an adder. It is probably confined to deciduous woods, especially associated with willow and alder, and is fairly common. The cap is very slimy when damp, shiny when dry and up to 10 cm across. The gills are first bluish, then rusty brown. The stem is light olive-coloured, brownish between the 'garlands', and can grow up to 12 cm high. The flesh is whitish to light yellow, brownish towards the base. The scent is slight and the taste mild, but this fungus should not be eaten. It grows from August to October.

Another large, grey-yellow to olive brown *Myxacium* species is *Cortinarius elatior* (☉) Fr. (from Latin *elatior*: higher). Its cap is first bell-like and then expanded. It is 5–10 cm wide, very slimy when damp and has a distinctly wrinkled edge. The stem is whitish with light violet bands, up to 15 cm tall. The flavour is mild and the fungus edible. It grows commonly during August and September in both coniferous and deciduous woods, especially beech.

Cortinarius collinitus ✪

Pers. ex Fr. (from Latin *collinitus*: greased)

This species has a pretty brownish yellow cap with a darker centre and a shining violet stem. It usually grows in mossy coniferous woodland, especially pine, but sometimes in deciduous woods such as beech. The convex cap has a low umbo and can grow to 10 cm wide. The gills are first grey-blue, then brown. The stem is of even thickness, up to 12 cm tall. The layer of violet slime sometimes dries to form violet bands. Found from August to October.

A related species is *Cortinarius mucosus* ✪ Fr. (from Latin *mucosus*: slimy) which is very slimy when damp. Its cap is entirely chestnut brown and the stem is almost white. Grows mostly in mossy pine woods on sandy soils, but is rare.

Cortinarius mucosus

Cortinarius trivialis

Cortinarius elatior

Cortinarius collinitus

Cortinarius armillatus

Fr. (from Latin *armillatus*: adorned with an armband)

Red-banded Cortinarius

The caps of many species have a skin which never becomes slimy or sticky. In certain cases in the subgenus *Hydrocybe* (from Greek *hydro*: water and *kybe*: head) the species change colour when moist and are hygrophanous. Some of the non-slimy species have a well-developed, prettily-coloured outer veil which often remains in the form of bands around the stem. These fungi belong to the subgenus *Telamonia* (from Greek *telamon*: broad linen bandage).

One species easily recognized by the pretty vermilion bands around its stem is *Cortinarius armillatus*. This is restricted to birch, and grows in acid soil, often on the edge of swamps. It is fairly common and widespread in Britain. The cap is fleshy, bell-shaped then expanding, smooth to delicately fibrous, red-brown and up to 12 cm across. The gills are broad, fairly widely spaced and cinnamon-coloured. The stem is red-brown, whitish and swollen towards the base, and grows to 12 cm high. The scent is slight and the taste mild, but this fungus should not be eaten because several species with bands around the stem are very poisonous (see below). It grows from August to October.

*Cortinarius
gentilis*

Cortinarius speciosissimus

Kühner & Romagnesi (from Latin *speciosissimus*: very pretty)

Cortinarius species have long been regarded as fairly harmless, even though many have an unpleasant taste. It was therefore something of a sensation in the 1950s when an orange-brown species, found growing in deciduous woods, was discovered to have caused some hundred cases of poisoning in Poland, twenty of them fatal. In the same group of red-brown species there are several which grow in coniferous woods, one of which, *C. speciosissimus*, has proved to be very poisonous. It is rarely found in Britain but is probably overlooked. Several other species within this group are also poisonous. The poisons take effect slowly and damage the kidneys and the liver, so these effects are not usually discovered for three to fourteen days.

This species grows among moss in pine woods. Its main features are the light cinnamon brown colour, pointed cap and gills, which are thick, one centimetre broad, widely spaced and ochre-coloured. The stem may have lemon yellow bands.

A related poisonous species which also grows in coniferous woods, is *Cortinarius gentilis* Fr. (from Latin *gentilis*: family). It also has thick and widely-spaced gills.

There is only one certain way to avoid fungus poisoning: stick to those fungi which are known to be edible.

Cortinarius armillatus

*Cortinarius
speciosissimus*

Cortinarius pholideus

Fr. (From Greek *pholis*: scale)

This pretty fawn-coloured species, with tiny scales on the cap
and often darker brown scales around the stem, is easy to
recognize. It is probably restricted to deciduous trees, especially birch. In Britain it is locally common, but our knowledge
of its conditions of growth and geographical distribution is
still very incomplete. This fungus belongs to the dry, non-
hygrophanous species, and was formerly placed in the
subgenus *Inoloma* (from Greek *is*: tree and *loma*: fringe,
border), a group of species with a fibrous or scaly cap and
often with a swollen base. These have now been divided
among other groups. *Cortinarius pholideus* has been placed
in the subgenus *Sericeocybe*. The cap is 3–6 cm wide and the
points of the scales turn upwards. The gills are at first
brownish yellow, often with a tinge of violet, and 5–10 cm
high. The flesh is pale brown. The scent is slight and the taste
mild, but this fungus – like most other brownish species –
should not be eaten. It grows during August and September.

Dermocybe cinnamomea

(L. ex Fr.) Wünsche (from Greek *derma*: skin and *kybe*: head,
plus *kinnamomon*: cinnamon)

Cinnamon Cortinarius

In the classic division of *Cortinarius* the subgenus *Dermocybe*
includes the non-hygrophanous species with a dry and silky
skin containing substances which colour them yellow,
greenish and red. The gills show particularly brilliant colours.
Chemical and other studies have shown that it is possible to
form a cohesive group of about twenty species. These now
form a genus of their own, *Dermocybe*, while other species
within the old group have been placed elsewhere. It is uncertain whether the *Dermocybe* species are mycorrhizal. Some
grow in marshes where there are no trees, while others grow in
deciduous and coniferous woods.

Here we illustrate three types, *Dermocybe cinnamomea*,
Dermocybe semisanguinea (Fr.) Wünsche (from Latin *semi*:
half and *sanguineus*: bloody) and *Dermocybe sanguinea*
(Wulf. ex Fr.) Wünsche. The names are descriptive of these
fungi. A number of other types can be identified by minor
points in the colouring of the gills, but there is much research
still to be done on their life cycle.

*Dermocybe
sanguinea*

96–97

*Cortinarius
pholideus*

*Dermocybe
cinnamomea*

*Dermocybe
semisanguinea*

Clitopilus prunulus ⊖

(Scop. ex Fr.) Kummer (from Greek *klitos*: bent and *pilos*:
cap, plus Latin *prunulus*: from Italian *prugnolo*: tricholoma)

The Miller

Pastures, mossy meadows, parks and gardens are the habitats
of numerous species with pinkish spore prints. One genus,
Clitopilus, differs in several ways from the other pink-spored
agarics. In particular, the gills are easily separated from the
cap. They are therefore thought to be related to *Paxillus* (see
vol. 1). The Miller is the most common, often growing in
groups or rings in woods and parks where there is plenty of
humus, but also in forests. It occurs throughout northern
Europe.

The cap, 5–10 cm across, is first grey and convex with an
in-turned margin, then grey-white and almost flat, with an
irregular wavy edge. The gills are dense and decurrent, first
white, then pinkish. The white stem grows to 6 cm high. The
flesh is white and the fungus is soft and brittle. The scent is
reminiscent of meal or compressed yeast and the taste is mild.
It makes delicious eating but can be confused with small white
Clitocybe species, some of which are poisonous. Another
possible source of confusion is the Livid Entoloma (see
below). The Miller grows from July to October.

Entoloma sinuatum ⊗

(Bull. ex Fr.) Kummer (from Latin *sinuatus*: wavy)

Livid Entoloma

The true pink-spored agarics, which have angular spores, are
placed in various genera. One of these is *Entoloma* (from Greek
ento: inner and *loma*: fringe), with the characteristics of
Tricholoma. The large species belong to this genus. The largest
of all is *Entoloma sinuatum*, which is poisonous. The cap,
5–15 cm across, is first matt light grey and convex with an inrolled margin. It then expands, becomes irregularly wavy
and turns grey-yellow and shiny. The gills are widely spaced
and sinuate, at first yellowish then pink because of the spores.
The stout stem is pale yellow and up to 12 cm high. The
scent suggests meal or rancid oil. This fungus grows in parks
and groves from July to September. When young it can be
confused with the St George's Mushroom or the Clouded
Agaric.

Most pink-spored species are small and inconspicuous.
Some, however, have a strong scent of meal, such as Silky
Nolanea *Nolanea sericea* ⊖ (Bull. ex Mérat) Orton (from
Latin *sericeus*: gleaming like silk). Others, like *Entoloma niti-
dum* ⊗ Quél. (from Latin *nitidus*: shining), have striking
colours.

*Nolanea
sericea*

*Entoloma
nitidum*

Clitopilus prunulus

Entoloma sinuatum

Lactarius vellereus ⊗

Fr. (from Latin *lactarius*: secreting milk, plus *vellereus*: woolly)

Milk caps and Russulas form a natural group of agarics, distinguished by their brittle, never fibrous, flesh in both cap and stem. The other main characteristic of the milk caps is an abundant supply of a seeping fluid which forms drops on damaged places, especially the gills. Many species are restricted to a specific kind of tree.

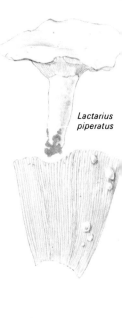

Lactarius piperatus

Lactarius vellereus is easily distinguished from other milk caps by the size, up to 25 cm across, the chalky white colour, the dry cap shaped like a shallow funnel and the widely-spaced gills. It grows mainly in deciduous woods, often in groups. In its early stages, the cap has an inrolled, rather woolly margin which later unrolls and has only a fine down until it becomes almost smooth. The gills are white, then yellowish or flecked with yellow. The thick white stem, 4–8 cm tall, is often tinged with lemon yellow and covered in fine down. The milky fluid is white and plentiful. There is a slight scent and an acrid taste. *Lactarius vellereus* is not edible. It grows in August and September.

A related species is *Lactarius piperatus* ⊝ Scop. ex Fr. (from Latin *piperatus*: peppered) with a somewhat smaller cap, 10–15 cm across, and a longer stem. The cap surface is smooth and dry and the gills are crowded. The colour is white with a tinge of yellow. The milky-coloured fluid is extremely hot but *Lactarius piperatus* is edible if well boiled. It grows during August and September in deciduous woods, especially in beech woods and among hazels.

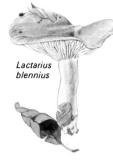

Lactarius blennius

Lactarius turpis ⊝

(Weinm.) Fr. (from Latin *turpis*: ugly)

Ugly Milk Cap

This dark species, which varies from greenish brown to almost black, grows throughout Britain under birch trees, especially on damp or peaty soils. In eastern Europe it is highly valued for salting. The cap is first greenish yellow, somewhat downy, with an inrolled margin, then turns darker and expands to 10–20 cm across. The crowded yellow gills turn dark grey when bruised. The stem is somewhat lighter than the cap, 3–5 cm high. The flesh is whitish grey. The milky fluid is white and the taste extremely hot, but this fungus is edible after boiling. It grows from July to October.

Another greenish species is the Slimy Milk Cap *Lactarius blennius* ⊗ Fr. (from Latin *blennius*: slimy). It has a greyish green cap, 3–6 cm broad, with darker concentric drop-like markings, grey-green gills and a paler stem. The white fluid is very peppery. This fungus is inedible. It grows in beech woods.

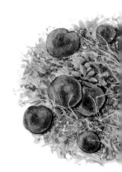

Lactarius vellereus

Lactarius turpis

Lactarius scrobiculatus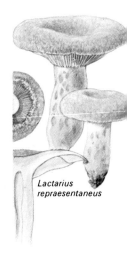

Scop. ex Fr. (from Latin *scrobiculatus*: pitted)

An important factor in identifying many milk caps is the fluid colour and the way it changes when exposed to air. Cut crossways into the gills with a knife to produce a good quantity of the fluid; in several species this turns from white to yellow. One of these is *Lactarius scrobiculatus*. Easy to recognize by its size, up to 20 cm across, its straw yellow colour and light yellow stem with darker orange spots, this fungus is particularly fond of damp coniferous woods, and, although rare in Britain, is common in northern Europe. It can be eaten after boiling and salting. At first the cap has an inrolled, downy margin but then expands and takes on the shape of a shallow funnel, sometimes with darker zones. It is 7–20 cm across. The gills are light yellow. The stem is 6–8 cm high and 2–3 cm thick. This fungus is sticky when damp. The scent is pleasant and the taste hot, but the acrid substances can be soaked away in water. *Lactarius scrobiculatus* grows in August and September.

Lactarius repraesentaneus

A species which closely resembles it is *Lactarius repraesentaneus* Britz. (from Latin *repraesentaneus*: representing). The sulphur yellow cap is 5–15 cm in width with an inrolled shaggy margin. The yellow stem grows to a height of 10 cm. The white fluid turns violet when exposed to the air. *Lactarius repraesentaneus* has a pleasant scent, and the taste is mild at first, becoming somewhat acrid. It grows in non-calcareous soil in both deciduous and coniferous woods, but in the British Isles is extremely rare outside Scotland.

Lactarius torminosus

Schaeff. ex Fr. (from Latin *torminosus*: suffering from colic)

Woolly Milk Cap

The Woolly Milk Cap occurs commonly in mixed woods and on heaths, especially on damp, non-calcareous soil. At first the cap has an inrolled woolly margin and then expands into a shallow funnel shape, becoming 6–12 cm in width. The cap surface is fibrous and pink with pale red zones. The gills are at first pale pink, then become tinged with yellow. The stem is also pale pink, often hollow and 4–8 cm high. The white fluid does not change colour. The scent is slight and the taste acrid. This fungus is edible after boiling; in Finland it is one of the most sought-after fungi for salting.

Lactarius pubescens

A related species, also with an acrid taste, is *Lactarius pubescens* Fr. (from Latin *pubescens*: with small hairs) which is smaller and paler in colour. It is only slightly woolly, never zoned, and has a shorter stem, often tapered towards the base. It grows in sandy soil under birches.

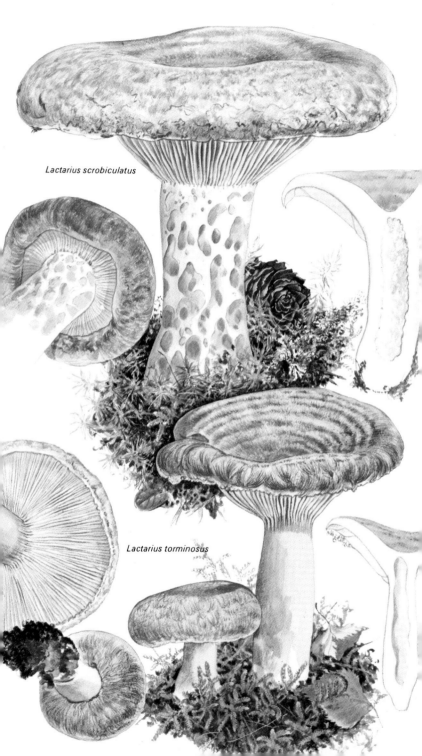

Lactarius scrobiculatus

Lactarius torminosus

Lactarius deliciosus ☉

(L. ex Fr.) S. F. Gray (from Latin *deliciosus*: delicious)

Saffron Milk Cap

The Swedish common-name *riska* is one of the oldest fungus names. It is derived from an Indo-European word which means to redden. The Saffron Milk Cap was described by Linnaeus and was also noted by one of his contemporaries, the early mycologist Jakob Schaeffer, who worked in southern Germany, as appearing in woods and on heaths during August. We know now that milk caps with orange or red fluid include quite a number of species, but they do not all taste good. They also have various habitats: some are restricted to pines, some to firs and some need chalky soil. The fleshy Saffron Milk Cap has a brick to grey-red cap with distinct darker, widely-spaced zones. The fairly short stem has orange 'pock-marks'. It grows near pines on open chalky land and at roadsides. The cap has an inrolled margin at first which then broadens into a shallow funnel shape. It is 5–15 cm broad, and sticky when damp. The crowded gills are somewhat decurrent and pale orange-yellow to orange. The stem, up to 7 cm high, is often of even thickness, but may taper towards the base. When cut, the light red flesh does not change colour for an hour or so. The scent is slight and the flavour pleasant. The Saffron Milk Cap frequently occurs in groups under conifers throughout northern Europe.

Lactarius deliciosus

Lactarius deterrimus ☉

Gröger (from Latin *deterrimus*: the worst)

This species is restricted to fir trees. It differs from the Saffron Milk Cap in that the cap has crowded narrow zones near the margin which are often barely perceptible, and there are no 'pock-marks' on the stem. After a few minutes the orange-coloured fluid turns wine red or purple. The fungus is mainly orange, but older specimens often have green patches, and frost turns them completely green. The cap, sticky when damp, is first convex with an inrolled margin and a somewhat depressed centre. Later it is shaped like a shallow funnel and is 10–15 cm in width. The flesh is thinner than the Saffron Milk Cap. The hollow stem grows up to 10 cm. The scent is slight and the taste slightly acrid, and this fungus is not as good to eat as the Saffron Milk Cap. It grows generally throughout Scandinavia in August and September, but it is not known in Britain. There are several related species in Europe.

Lactarius deterrimus

Lactarius rufus

(Scop. ex Fr.) Fr. (from Latin *rufus*: red-brown)

Rufous Milk Cap

There are sixty milk cap species in Britain. Several of them are coloured in shades of brown, including *Lactarius rufus* and *Lactarius helvus*. Both are very common in coniferous woods on non-calcareous soil throughout Britain and Europe. *Lactarius rufus* forms mycorrhizal associations with pines, but is not strictly confined to them as it can also be found under birches and among heather.

This fungus is easy to recognize by the glistening red-brown cap with the conical pointed umbo, and by the hot taste. The cap is first convex with an inrolled margin, then funnel-shaped, dry and 5–10 cm wide. The gills are first pink, becoming light yellow-red. The stem is red-brown, often hollow, and up to 8 cm high. The flesh is white. The fluid is white and does not change colour. The scent is slight, and after the acrid substances are soaked out in water the fungus can be eaten, but opinions about its merits differ. In eastern Europe it is salted in barrels and thought to be a tasty dish when fried with bacon. The Rufous Milk Cap grows from August to October.

Lactarius helvus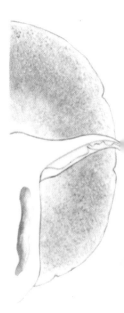

Fr. (from Latin *helvus*: honey yellow)

This milk cap is particularly fond of wet places. It occurs under or near conifers but is only found occasionally. The cap, which can grow to 10 cm in width, distinguishes it from the smaller fungi found in similar habitats.

Lactarius helvus is grey-yellow to light yellow-brown and the cap surface varies from downy to rough. The fluid is as clear as water. The flat cap with an inrolled margin and a slightly umbonate centre develops into a shallow funnel shape, 6–15 cm in width. The crowded gills are pale yellow, but with age take on the colour of the cap. The yellow-brown stem is hollow, and up to 12 cm high. The flesh is yellow-brown. The unique scent, especially when the fungus is dried, suggests liquorice or dried packet soup. *Lactarius helvus* has a mild taste, but should only be eaten after boiling. It is one of the few milk caps that are undoubtedly poisonous, although only mildly so. When dried it can be used in small quantities as a spice. Its growing season is from August to October.

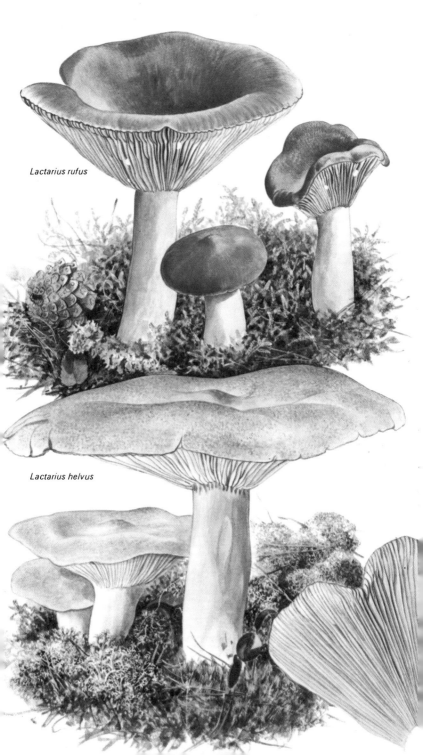

Lactarius rufus

Lactarius helvus

Lactarius volemus

(Fr.) Fr. (from Latin *volema pira*: a Red Warden pear)

In Sweden, a number of milk caps were collectively called *Brötlingar*, a word derived from the German *Brot* meaning bread. But as most of the species are acrid, the name refers to *Lactarius volemus*, which has a mild taste and was eaten raw in certain parts of Germany.

Lactarius volemus is large and sturdy, and is easily recognized by the yellow-brown to orange colour, yellow-white gills which turn brown when touched and abundant white fluid. It has two forms of colouring: the red-brown type grows in mossy coniferous woods and beech woods, and the fiery yellow kind is found only in woods of beech and oak. The cap is at first convex, then develops into the shape of a shallow funnel. It is never sticky, and grows to 5–15 cm across. The gills are crowded and the stem is stout, up to 12 cm high, and is the same colour or somewhat lighter than the cap. The firm, pale yellow flesh gradually acquires patches of brown. The scent suggests boiled shellfish and the taste is reminiscent of almond. This fungus is excellent to eat when thoroughly fried. It can also be eaten raw with salt. *Lactarius volemus* grows in August and September.

Lactarius mitissimus

(Fr.) Fr. (from Latin *mitissimus*: very mild)

This fungus resembles a small form of *Lactarius volemus*. It has a pretty, brown-yellow circular cap, somewhat shiny and with no zones. The white fluid is mild at first with an acrid after-taste. *Lactarius mitissimus* is found in both deciduous and coniferous woods, but mainly it grows close to hazel, oak and maple. Elias Fries noted that one form found in beech woods is extremely mild. Reports that this species grows mainly in mountainous pine woods in central Europe suggest that there may be several related species or forms. The small brownish milk caps are very numerous, and not all of them have been sufficiently investigated. There may be as many as twenty different species. The cap of *Lactarius mitissimus* is 2–6 cm in width and has pale yellow to pale orange-yellow gills. The stem is brownish yellow and firm, of even thickness, 3–8 cm tall. The flesh is yellowish. The scent is slight, and the species is edible, but forms with a bitter taste should be boiled. It is fairly common between August and October.

*Lactarius
volemus*

Lactarius mitissimus

Lactarius flexuosus

(Pers. ex Fr.) S. F. Gray (from Latin *flexuosus*: with many bends)

At least ten European milk caps are greyish, but some vary from blue to violet and are among the most common woodland fungi. One of the grey species is *Lactarius flexuosus* which is not uncommon among grass and moss in woodland glades, copses and beside roads. It is easily recognized by the often irregular, wavy, fleshy, violet-grey cap, pale yellow, widely-spaced gills and short compact stem. The cap, 6–15 cm wide, may be somewhat zoned. The pale blue-grey stem, 3–6 cm tall and 1.5–2 cm thick, tapers towards the base. The white fluid does not change colour. The fungus has a slight scent and an acrid flavour but can be eaten after boiling. It grows from August to October.

Lactarius vietus Fr. (from Latin *vietus*: wilted, withered) is a smaller pale grey to brown-grey species with fluid that turns grey when dry. It has a cap 3–8 cm broad with pale yellow, widely-spaced gills which develop dark grey spots when injured. The stem is whitish to light grey, up to 8 cm tall. The scent is slight and the taste ranges from mild to rather acrid. *Lactarius vietus* grows in large groups on damp ground under birches. It is very common throughout Britain and Europe from August until the first frosts.

Lactarius vietus

Lactarius trivialis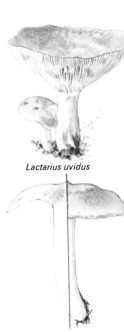

(Fr. ex Fr.) Fr. (from Latin *trivialis*: ordinary, common)

This large, glutinous milk cap is an uncommon species found on wet mossy ground. It occurs in two forms: grey-lilac and yellow-grey. In eastern Europe it is highly valued for salting and is often seen in Finnish markets. The cap at first has an inrolled margin which later expands to develop a slightly depressed centre. It is up to 15 cm in width and is often faintly zoned. The stem, 6–18 cm high, is paler than the cap, hollow, and may look somewhat swollen. The fluid is white. The scent is slight and the taste acrid. *Lactarius trivialis* is edible but good to eat only if boiled. It grows from August to October.

Another grey-violet milk cap is the middle-sized (4–8 cm) species *Lactarius uvidus* Fr. (from Latin *uvidus*: damp, wine-moistened). It is often sticky and has pale-coloured gills and white fluid. The white fluid, like the flesh, turns violet when exposed to the air. It grows under birches but is rare.

Lactarius uvidus

110–111

Lactarius flexuosus

Lactarius trivialis

Russulas

Some of the most conspicuous fungi in August and September are the brightly-coloured Russulas. They are often ignored because it is difficult to distinguish between the various species. This is a pity, because most mild-tasting Russulas are edible, and many are delicious. With the milk caps they form a natural group of fungi (see p. 100) with brittle flesh rather like a ripe apple. Russulas, however, do not have a milky fluid. In order to identify them one must study the colour of the spore print and cap, and the scent and taste of the flesh. Frequently a microscope and chemical tests are needed to tell the species apart. On the following pages we have grouped together those species of roughly the same colour.

Russula delica ⊗

Fr. (from Latin *russulus*: reddish [a large number of species are red] and *delicus*: weaned)

This whitish Russula belongs, like the others shown on these two pages, to a characteristic group of species with firm flesh, thick gills and more or less black and white colourings. None of these fungi is good to eat. The *Russula delica* cap is 5–15 cm wide and the white decurrent gills have a slight blue-green tinge, flecked with brown when the fungus is older. The white stem is about 5 cm tall. The distinctive scent is reminiscent of fish and the taste is acrid, especially in the gills. The fruit-body develops on the surface of the soil, and the cap is often covered with bits of grass, pine needles and leaves. This common fungus grows between July and October in both deciduous and coniferous woods.

The caps of Russula integra (*below*) *and* Russula adusta *can be very similar in colour*

Russula adusta ⊗

(Pers. ex Fr.) Fr. (from Latin *adustus*: blackened, scorched)

The cap of this species is first grey-white, then grey-brown, 5–15 cm across. The gills are decurrent and crowded and the stem is stout. The flesh becomes grey to red-brown. The scent suggests old red wine or wine-casks, and the flavour is almost mild. This species is uncommon in Britain, occurring mainly in pine woods.

A darker species – white to greyish, then almost black – is *Russula albonigra* ⊗ (Krombh.) Fr. (from Latin *albus*: white and *niger*: black). The flesh is white at first but becomes grey-black when broken. The taste is mild or slightly acrid. This species, which at certain stages of its development may be hard to distinguish from the one described above, is an uncommon fungus found in mixed woods from July to October.

Russula albonigra

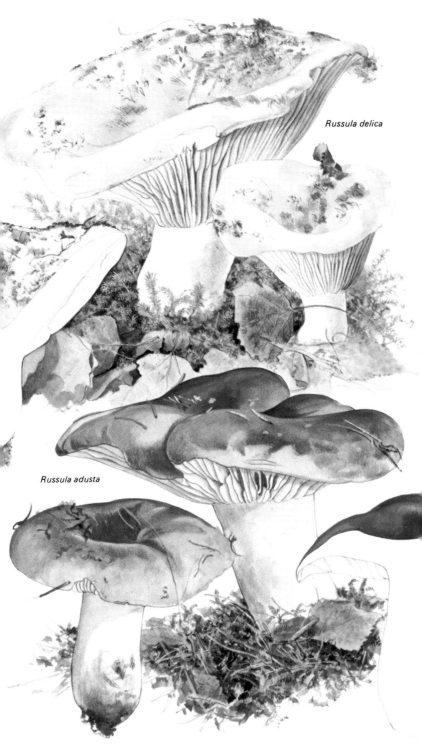

Russula delica

Russula adusta

Russula claroflava ⊖

Grove (from Latin *clarus*: clear, shining and *flavus*: light yellow)

The most striking-looking fungi are those which either spring up in great quantity or are bright yellow, orange or red like the Russulas. Several of the Russulas have an acrid taste and are more or less inedible, but the yellow and brick-red ones, with flesh which turns grey, have a mild taste and are good to eat. One of the common yellow species is *Russula claroflava*, whose cap is 5–12 cm wide and somewhat slimy when damp. It is brilliant yellow to buttercup yellow but older specimens become paler. The cap often has a depressed centre and the skin peels off easily. The white gills gradually turn creamy yellow as the spores mature. The white stem is 5–10 cm high and then turns grey. The scent is slight and the taste mild. This is an edible species, occurring generally during August and September in mixed deciduous woods, usually under birches or aspens. It is widespread throughout northern Europe.

It is easy to confuse *Russula claroflava* with the Common Yellow Russula *Russula ochroleuca* ⊖ (Pers.) Fr. (from Greek *ochros*: ochre and *leukos*: white). This has a somewhat acrid taste. The cap is roughly the same colour as *Russula claroflava* although with more ochre and a slight green tinge. The stem turns slightly grey in older specimens, but not as markedly as with the species described above. The Common Yellow Russula is abundant in deciduous and coniferous woods throughout Britain.

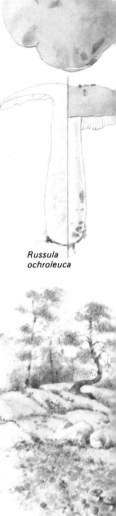

Russula ochroleuca

Russula decolorans ⊖

(Fr.) Fr. (from Latin *decolorans*: losing its colour)

One of the few Russulas occasionally to be found on barren moorlands where there are some pine trees and plenty of lichens. The red-yellow to brick red cap has an inrolled margin when young. It is 8–12 cm across with a somewhat depressed centre. The stem, which can grow to a good 10 cm, is at first white, becoming grey to dark grey. The flesh turns grey when broken, and older specimens are grey to dark grey both inside and out. The scent is slight and the taste mild. It is good to eat, but only young, white and fairly firm specimens should be used. The species is uncommon in Britain, and perhaps restricted to Scotland.

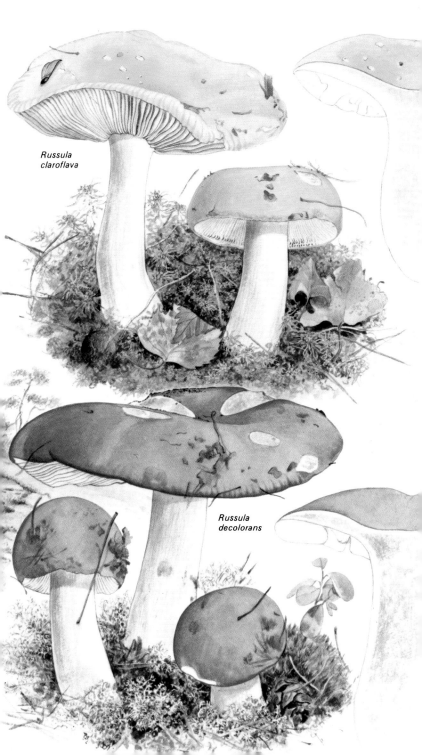

*Russula
claroflava*

*Russula
decolorans*

Russula puellaris

Fr. (from Latin *puellaris*: girlish, youthful)

This small delicate fungus can also be included among the yellow and brown Russulas because it turns increasingly yellowish with age. When young, however, it always has touches of several colours, particularly sienna yellow, and varying shades of yellow to light green, light red and light brown or violet. The thin cap is 3–6 cm wide and the gills are creamy yellow. The streaky, dirty, yellow stem is 3–5 cm tall and the flesh is brittle. It has a mild taste and no characteristic scent. It is edible and tastes good. This fungus occurs fairly frequently with both deciduous and coniferous trees.

Russula integra (*above*) *may be confused with* Russula puellaris

It is possible to confuse this species with *Russula versicolor* ⊖ J. Schaeff. (from Latin *versicolor*: changing colour). The latter, however, always has a tinge of grey in its other, very varying colours. The usually acrid taste also makes it easy to distinguish. It occurs frequently under birches. Small, light brown or brownish yellow forms of *Russula integra* can also be confused with this species.

Russula foetens ⊗

(Pers. ex Fr.) Fr. (from Latin *foetens*: evil-smelling)

Foetid Russula

The Foetid Russula can be recognized by the rank and revoltingly rancid smell. The fleshy, slimy cap and the brown-spotted gills also make an instantly repulsive impression. *Russula foetens*, which is believed to be poisonous, often grows in groups in deciduous woods, pastures and parks during summer and early autumn, and is often attacked by larvae, snails and parasitic fungi. The dirty yellow to yellow-brown cap, 8–15 cm across, has a strongly tuberculate and furrowed margin. The stem grows to a height of 5–10 cm and gradually splits apart inside to create a 'staircase' effect. The taste is hot and burning. The species is common and widespread throughout Britain and Europe. *Russula laurocerasi* ⊗ Melzer (from *laurus cerasus*: cherry laurel) is a less common but very similar species, distinguished by the distinctive odour of marzipan.

Russula
puellaris

Russula
foetens

Russula paludosa ⊖

Britz. (from Latin *paludosus*: marshy, swampy)

The majority of Russulas have some shade of red in the cap; those with bright red markings usually have an acrid taste and are not normally eaten raw. The largest of the red species is *Russula paludosa*. The cap can grow up to 20 cm in width, and when dry it is a shining apple red. Older specimens are often brownish red. The white gills turn light creamy yellow. The white, sometimes pinkish, stem is 10–15 cm tall. Young specimens have a slightly acrid taste. There is some disagreement about the merits of *Russula paludosa* for the table. In Finland it is picked and sold commercially. At one time it was thought that it should be boiled before eating, but this does not, in fact, seem to be necessary. This fungus is uncommon in Britain, possibly restricted to Scotland, and grows in damp coniferous woods, especially at the edge of swamps and in peat moss.

Russula lepida ⊗

Fr. (from Latin *lepidus*: dainty, pretty)

Russula emetica

This Russula flourishes in deciduous woods, especially under beech and oak. The thick firm cap is 5–10 cm in width and looks as if it had been powdered white. It is vermilion to carmine or orange-red. The white to creamy yellow gills often have a reddish edge. The stem is 4–10 cm high, rose-coloured or vermilion towards the bottom, sometimes flecked with yellow. The flesh tastes bitter. This fungus is inedible.

The two Russulas we have described can sometimes be confused with the emetic Russula *Russula emetica* ⊗ (Schaeff. ex Fr.) S. F. Gray (from Greek *emetica*: causing sickness) and *Russula sanguinea* ⊗ Bull. ex Fr. (from Latin *sanguis*: blood). The emetic Russula has been divided into several varieties. The typical form is a relatively rare fungus which grows in damp woods, on moss and at the edge of swamps. It is light red to bright cherry red with white gills and a white stem. The soft fragile flesh has a very acrid taste.

Russula sanguinea

Russula sanguinea has a blood red to pale red cap and pale yellow, somewhat decurrent gills. The white stem has a touch of red. The scent is fruity and the taste slightly acrid. It is an occasional species of damp pine woods during summer and autumn.

Another pretty Russula, resplendent in yellows and reds, is *Russula aurata* ⊖ With. ex Fr. (from Latin *auratus*: gilded, decked in gold). An uncommon species, it grows under beeches, oaks and hazels. The entire fungus is tinged with golden yellow. It has a mild taste and is good to eat.

Russula paludosa

Russula aurata

Russula lepida

Russula integra ⊖

Fr. (from Latin *integer*: whole, pure, sound, healthy)

The most important Russula species from the cook's point of view, and a good edible fungus, it grows in deciduous woods, but is unfortunately rare. The fruitbody often develops on the surface of the soil, and young fruitbodies may be completely hidden by pine needles, twigs and leaves. The cap is 6–12 cm in width and is convex at first, then expands and becomes centrally depressed. The cap skin is sticky and glistening when damp, with colours varying widely from brownish yellow to chocolate brown, brown-violet or purple-red. This variation occurs partly because the colouring substances of Russulas are soluble in water. After rain the caps can look completely bleached. The thick and widely-spaced gills have low transverse ridges, especially at the cap edge. They finally become deep straw yellow. The stem grows to 10 cm tall, with a firm outer layer which is porous internally. The scent is mild and the flavour suggests fresh nuts or almonds. This fungus can be eaten raw in salad but is also tasty when fried or cooked in sauce. It usually grows during August and early September.

Russula vesca ⊖

Fr. (from Latin *vescus*: edible)

This species is very like certain forms of *Russula integra* in colour, but it can be recognized by the cuticle that does not reach all the way to the cap edge. In full-grown specimens the flesh of the cap and the gills form a white margin. The cap is 6–12 cm broad, vinaceous pink with a tinge of brown, or

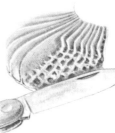

Russula integra *gills are connected by transverse ridges*

brownish and flecked with yellow-brown. The white gills are speckled with rusty brown, and some of them fork. The stem is white above, flecked with yellow or rust brown lower down. The firm flesh has a sweetish smell, reminiscent of fruit. The taste is mild and nutty. *Russula vesca* is excellent to eat, and is a common species of deciduous woods, especially beech and oak, from July to October.

It is possible to confuse these species with certain acrid Russulas, for example *Russula badia* ⊖ Quél. (from Latin *badius*: chestnut brown). This has a red-brown to red-black matt surface and very often the stem has a reddish tinge towards the base. The thin gills are crowded but of the same ochre to egg yellow as those of *Russula integra*. The scent suggests cedarwood oil, the taste slowly becomes very acrid and the species is inedible. It occurs in autumn and is a rare species of coniferous woods in Scotland.

Russula badia

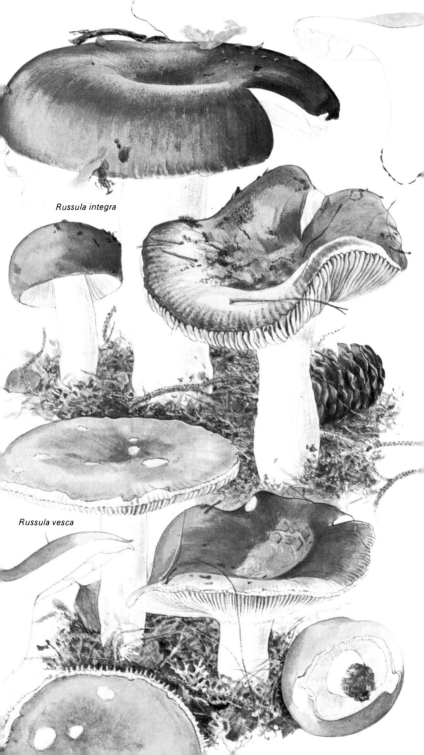

Russula integra

Russula vesca

Russula xerampelina ⊖

(Schaeff. ex Secr.) Fr. (from Greek *xerampelinos*: the colour of dry vine-leaves)

The Latin name is a useful description of this Russula, considered one of the best of all fungi for the table. It has many colour variations and is sometimes divided into species or sub-species. The dry matt cap is 5–15 cm across. In some forms it is carmine, purple-red, wine red or blackish red, while others are red-brown, olive green or olive brown, often with traces of yellow. The pale yellow gills are sometimes more ochre. The white stem, 4–6 cm high, is tinged with red, especially at the base. The gills and flesh stain or turn dirty brown when bruised or broken. It smells of crab or boiled shellfish and the taste is mild; all dishes made with this fungus have a more or less pronounced shellfish taste. Frequently found from August to October in deciduous woods, especially under beech and oak, but may also occur in coniferous woods. Bright red forms can be confused with the acrid-tasting emetic Russula. This, however, always has pure white gills and a white stem.

Russula obscura ⊖

Romell (from Latin *obscurus*: dark)

This fairly large Russula grows in coniferous woods among lichens and mosses and is another of the more or less red to violet-red species. The cap is firm, 6–12 cm in width, sticky in damp weather and either shiny or matt when dry. The somewhat sulcate margin often looks as if it has been powdered white. The colour is wine red to purple-brown, with tinges of yellow and orange in the centre. The gills are pale to creamy yellow. The white stem, 5–10 cm tall, becomes grey with age, as does the entire fungus. Even in young specimens the basal mycelium is grey, and in older ones the stem may become entirely black. The scent is rather honey-like in specimens that have been left for awhile. The taste is mild, and this species is good to eat.

Russula queletii

Another of the red and violet species is *Russula queletii* ⊗ Fr. (from the name of the French scholar Quélet). It may be carmine to pink, with a tinge of violet, or may be pure violet. The stem is light red or white. The scent is fruity, suggesting stewed gooseberries, and the taste is acrid. This uncommon species grows in damp places in coniferous woods.

Russula xerampelina

Russula obscura

Russula aeruginea ⊖

Lindblad ex Fr. (from Latin *aerugineus*: verdigris)

Grass-green Russula

The Grass-green Russula is found in clumps of birch and aspen, and in pastures under birches. A similar species, which grows mainly under oaks and hazels, and is often more variable in colour, is *Russula urens* ⊖ Romell *apud* J. Schaeff. (from Latin *urens*: burning). It is inedible, but the Grass-green Russula is good to eat. *R. urens* is not found in Britain. The *Russula aeruginea* cap is 5–10 cm across with a furrowed margin. It is sticky when damp and varies from grey-green to verdigris green or sometimes blue-green to blackish green. It sometimes resembles the green forms of *Russula xerampelina* (see illustration on this page). The gills are creamy yellow and the white stem with firm flesh grows 4–6 cm tall. The scent is slight, and the mild or slightly acrid taste vanishes when the fungus is cooked. It is frequently found throughout the British Isles between July and October.

Green form of
Russula xerampelina

Russula cyanoxantha ⊖

(Schaeff. ex Secr.) Fr. (from Greek *kyanos*: dark blue and *xanthos*: yellow)

The habitats of this variously-coloured species, which grows up to 15 cm in width, are parks, meadows and deciduous woods. It is very common throughout Britain and in some years fruits in great numbers. It is sometimes called the Parrot Russula in central Europe because of its cap colours. Young fruitbodies are slate grey to grey-violet, but older specimens become violet to blue-violet with more or less pronounced touches of green. The pale white gills are slightly decurrent and usually variously forked. The white stem, 5–10 cm tall, is tinged with violet towards the base. There is no particular scent, but it has a mild taste and makes a fine addition to a mixed mushroom dish.

One of the more pure blue or violet species is *Russula azurea* ⊖ Bres. (from Latin *azureus*: azure blue). It is fairly small with a blue-grey cap 3–8 cm across, white gills and a white stem. It is mild-flavoured and edible. *Russula azurea* is confined to coniferous woods and is very rare throughout northern Europe. A similar species, *Russula parazurea* ⊖ J. Schaeff. (from Greek *para*: near, and Latin *azureus*), grows under beech and oak.

Russula azurea

*Russula
aeruginea*

*Russula
cyanoxantha*

Bibliography

General and Reference Works on Fungi

Ainsworth, G. C., *Introduction to the History of Mycology*, Cambridge University Press, 1976.

Ainsworth, G. C., and Bisby, G. R., *A Dictionary of the Fungi*, Commonwealth Mycological Institute, 1971.

Ainsworth, G. C. and Sussman, A. S. (eds), *The Fungi: An Advanced Treatise*, Vol. 1, *The Fungal Cell*, 1965, Vol. 2, *The Fungal Organism*, 1966, Vol. 3, *The Fungal Population; Ecology*, New York and London, Academic Press, 1968.

Ainsworth, G. C., Sparrow, F. K. and Sussman, A. S. (eds), *The Fungi: An Advanced Treatise*, Vol. 4A, 1973, *Ascomycetes and Fungi Imperfecti*, Vol. 4B, 1973, *Basidiomycetes and Lower Fungi*, New York and London, Academic Press, 1974.

Alexopoulos, C. J., *Introductory Mycology*, 2nd edn, New York, John Wiley & Sons, 1962.

Burnett, J. E., *Fundamentals of Mycology*, Edward Arnold, 1968.

Cartwright, K. St G. and Findlay, W. P. K., *Decay of Timber and Its Prevention*, 2nd edn, H.M.S.O., 1958.

Christiansen, C. M., *The Molds and Man. An Introduction to the Fungi*, 3rd edn, University of Minnesota, 1965.

Fries, E. M., *Systema Mycologicum*, Vols. 1–2, Lundae, 1821–23, Vol. 3 and Index, Gryphiswaldae, 1829–32, (1821–32).

Haas, H., *The Young Specialist Looks at Fungi*, Burke Publishing Company, 1969.

Hawker, L. E., *Fungi. An Introduction*, 2nd edn, Hutchinson University Library, 1974.

Heim, R., *Champignons toxiques et hallucinogènes*, Paris, N. Boubée & Cie, 1963.

Ingold, C. T., *The Biology of Fungi*, 3rd edn, Hutchinson Educational, 1973.

Large, E. C., *The Advance of the Fungi*, Cape, 1940.

The British fungus flora

Ainsworth, G. C. and Sampson, K., *The British Smut Fungi* (Ustilaginales), Commonwealth Mycological Institute, 1950.

Barnett, H. L. and Hunter, B. B., *Illustrated Genera of Imperfect Fungi*, Minneapolis, Burgess Publishing Company, 1972.

Bourdot, H. and Galzin, A., *Hyménomycètes de France Hetérobasidiés – Homobasidiés gymnocarpes*, Bibliotheca Mycologica 23, Paris, J. Cramer, 1928.

Brightman, F. and Nicholson, B. E., *The Oxford Book of Flowerless Plants*, Oxford University Press, 1966.

Christiansen, M. P., *Danish Resupinate Fungi*, Part 1, *Ascomycetes and Heterobasidiomycetes*, Part 2, *Homobasidiomycetes*, Dansk Botanisk Arkiv 19, 1959, 7–309.

Corner, E. J. H., *A Monograph of Clavaria and Allied Genera*, Oxford University Press, 1950; supplementary volume. Beihefte zur Nova Hedwigia 33, 1970.

Dennis, R. W. G., *British Ascomycetes*, Cramer, 1968.

Dennis, R. W. G., Orton, P. D. and Hora, F. B., *New Check List of British Agarics and Boleti*, Supplement to Transactions of the British Mycological Society 43, 1960.

Domanski, S., *Fungi, Polyporaceae I (resupinatae), Mucronoporaceae I (resupinatae)*, translated from Polish; available from US Department of Commerce, Springfield, Virginia, 1972.

Domanski, S., Orlos, H. and Skirgiello, A., *Fungi, Polyporaceae II (pileatae), Mucronoporaceae II (pileatae), Ganodermataceae, Bondartzewiaceae, Boletopsidaceae, Fistulinaceae*, 1973; available as *Polyporaceae I*.

Eckblad, F. E., *The Gasteromycetes of Norway*, Nytt magasin for botanikk, 19–86, 1955.

Ellis, E. A., *British Fungi*, Book 1, *Larger species*; Book 2, *Smaller Species*, Norwich, Jarrold Colour Publications, 1976.

Eriksson, J. and Ryvarden, L., The *Corticiaceae of North Europe*, Vol. 2, *Aleurodiscus to Confertobasidium*, Vol. 3, *Coronicium to Hyphoderma*, 1975, *Hyphodermella to Mycoacia*, 1976; Vol. 1, with key to genera and glossary, will be published as a final volume; Norway, Fungiflora.

Findlay, W. P. K., *Wayside and Woodland Fungi*, Frederick Warne & Co., 1967.

Grove, W. B., *British Stem- and Leaf-Fungi (Coelomycetes)*, Vol. 1, *Sphaeropsidales*, Vol. 2, *Sphaeropsidales and Melanconiales*, Cambridge University Press, 1935–7; reprint 1967.

Henderson, D. M., Orton, P. D. and Watling. R., *British Fungus Flora. Introduction to Families and Genera*, Edinburgh, H.M.S.O., 1969.

Ing, B., *A Census Catalogue of British Myxomycetes*, The Foray Committee of the British Mycological Society, 1968.

Ingold, C. T., *An Illustrated Guide to Aquatic and Water-borne Hyphomycetes (Fungi Imperfecti)*, Freshwater Biological Association, Scientific Publication No. 30, 1975.

Kuhner, R. and Romagnesi, H., *Flore analytique des champignons supérieurs*, Paris, Masson et Cie, 1953; reprint 1974.

Lange, M. and Hora, F. B., *Collins Guide to Mushrooms and Toadstools*, 1963.

Martin, G. W. and Alexopoulos, C. J., *The Myxomycetes*, University of Iowa Press, 1969.

Moser, M., *Kleine Kryptogamenflora* edit. H. Gams), Vol. 2A, Ascomyceten, Stuttgart, Gustav Fischer Verlag, 1963.

Moser, M., *Kleine Kryptogamenflora* Vol. 2B, *Basidiomyceten II, Die Rohrlinge und Blätterpilze*, 3rd edn, Stuttgart, Gustav Fischer Verlag, 1967.

Pegler, D. N., *The Polypores*, Bulletin of the British Mycological Society 7: Supplement, 1973.

Ramsbottom, J., *A Handbook of the Larger British Fungi*, British Museum (Natural History), 1923; reprint 1965. J. Cramer.

Ramsbottom, J., *Mushrooms and Toadstools*, Collins, 1953.

Rayner, R. W., *Keys to the British Species of Russula*, Bulletin of the British Mycological Society, Vol. 2–4, 1968–70.

Reid, D. A., *A Monograph of the British Dacrymycetales*, Transactions of the British Mycological Society, 62, 433–94.

Richardson, M. J. and Watling, R., *Keys to Fungi on Dung*, reprinted from the Bulletin of the British Mycological Society, 2, 1968, 18–43; 2nd edn, 1975.

Rinaldi, A. and Tyndalo, V., *Mushrooms and Other Fungi* – an Illustrated Guide, Hamlyn, 1974.

Ryvarden, L., *The Polyporaceae of North Europe*, Vol. 1, *Albatrellus to Incrustoporia*, Norway, Fungiflora, 1975.

Singer, R., *Mushrooms and Truffles. Botany, Cultivation and Utilization*, Leonard Hill Books, 1961.

Singer, R., *The Agaricales in Modern Taxonomy*, 3rd edn, Vaduz, J. Cramer, 1975.

Wakefield, E. M. and Dennis, R. W. G., *Common British Fungi*, Gawthorn, 1950.

Watling, R., *British Fungus Flora 1. Boletaceae, Gomphidiaceae, Paxillaceae*, Edinburgh, H.M.S.O., 1970.

Watling, R., *Identification of the Larger Fungi*, Hulton Educational Publications, 1973.

Webster, J., *Introduction to Fungi*, Cambridge University Press, 1970.

Wilson, M. and Henderson, D. M., *British Rust Fungi*, Cambridge University Press, 1966.

For additional and more detailed references, the reader should consult the following list:

Holden, M., *Guide to the literature for the identification of British Fungi*, 3rd edn, Bulletin of the British Mycological Society, 9, 1975, 67–106.

Index of common English names

Index of Latin names

130